TRANSPORT OF MULTIPLE *ESCHERICHIA COLI* STRAINS IN SATURATED POROUS MEDIA

T0303907

TRANSPORT OF MULTIPLE *ESCHERICHIA COLI* STRAINS IN SATURATED POROUS MEDIA

DISSERTATION
Submitted in fulfilment of the requirement of
the Board for Doctorates of Delft University of Technology
and of
the Academic Board of the UNESCO-IHE Institute for Water Education
for the Degree of DOCTOR
to be defended in public,
on Tuesday 12 January, 2012, at 15:00 hours
in Delft, The Netherlands

by

George LUTTERODT

Master of Science in Environmental Engineering and Sustainable Infrastructure,
Royal Institute of Technology (KTH), Sweden
born in Osu, Accra, Ghana

This dissertation has been approved by the Supervisor:
Prof. dr. S. Uhlenbrook

Composition of the Doctoral Committee:

Chairman	Rector Magnificus Delft University of Technology
Vice-Chairman	Rector UNESCO-IHE
Prof.dr. S. Uhlenbrook	UNESCO-IHE / Delft University of Technology, supervisor
Prof.dr. F. Kansiime	Makerere University, Kampala, Uganda
Prof.dr.dr.h.c.ir.M.C.M.van Loosdrecht	Delft University of Technology
Prof.dr. G.J. Medema	TU Delft KWR / Water Cycle Research Institute
Prof.dr.P.N.L. Lens	UNESCO-IHE / Wageningen University
Dr. J.W.A. Foppen	UNESCO-IHE
Prof.dr. D. Brdjanovic	UNESCO-IHE / Delft University of Technology, reserve member

CRC Press/Balkema is an imprint of the Taylor & Francis Group, an informa business

© 2012, George Lutterodt

Published by:
CRC Press/Balkema
PO Box 447, 2300 AK Leiden, the Netherlands
e-mail: Pub.NL@taylorandfrancis.com
www.crcpress.com - www.taylorandfrancis.co.uk - www.ba.balkema.nl

ISBN 978-0-415-62104-5 (Taylor & Francis Group)

Acknowledgements

God makes things beautiful in his own time, this work got to this stage through mental, physical and emotional strength provided by God Almighty. I will like to express my sincere gratitude to all who in various ways contributed to the making of this thesis, and would like to state that many people directly or indirectly contributed to this thesis and those whose names are not mentioned here are the very good friends who would not need their names to be mentioned here to show appreciation to their contribution towards this thesis.

I am most grateful to my promoter Professor Stefan Uhlenbrook and supervisor Dr. Jan-Willem Foppen for their help and guidance. They made this thesis a reality. My profound appreciation to Dr. Wilfred Rölling of the Vrije University, Amsterdam for offering me the opportunity to do part of this work in his laboratory. Special acknowledgements goes to the entire staff of the UNESCO-IHE laboratory, they tirelessly worked behind the scenes to ensure successful running of experiments. I will also like to offer sincere appreciation to Eric Oduroh-Boahene of Geogroup Limited in Kumasi, Ghana, and Michael Nii Armah Aryeetey of the Ghana National Petroluem Corporation for continuously reminding me of the importance of this work and encouraging me. Special thanks goes to Nii Boi Ayibotele, Chief Consultant of Nii Consult, Dr. Stephen Dapaah-Siakwan (May his soul rest in peace) formerly of Water Research Institute of the Council for Scientific and Industrial Research (WRI-CSIR) in Accra, Ghana and Professor Bo Olofsson of the Department of Land and Water Resources Engineering of the Royal Institute of Technology (KTH), Stockholm. I would also want to thank my colleague and friend Henry Osei of Guldrest Resources Company Limited, Professor David Atta-Peters and Dr. Johnson Manu both of the Department of Earth Science at the University of Ghana, Legon, for their encouragement. I would want to show extreme appreciation to my colleagues and friends from Uganda; Alex Katukiza, Philip Nyenje and John Bosco Isunju for the good times we shared both in Delft and in Kampala. Through the study I made some good friends who were very supportive in various ways contributing directly or indirectly to the thesis, I will like to acknowledge Dr. Sharon Walker of the University of Carlifornia, Riverside, USA for her continuous encouragement, Mohan Basnet and Ahmed Maksoud both of whom I had the privilege of working with during their MSc and also Olivier Aprin. Special thanks goes to PhD fellowship officer Jolanda Boots. I would like to appreciate the time spent with the Ghanaian community at UNESCO-IHE and with Richard Quansah, a Ghanaian resident in Delft.

I am grateful to the Netherlands fellowship Program of NUFFIC for the award of a fellowship which made this study possible. I am also grateful to the Netherlands Ministry of Development Cooperation (DGIS), who funded part of this research through the UNESCO-IHE Partnership Research Fund.

I wish to express my sincere gratitude to Dr. Robinah Kulabako of the Public Health and Environmental Engineering Department of the Faculty of Technology, University of Makerere, Uganda. Appreciation also goes to Peter Kiyaga, who was so helpful during the field work in Kampala. Special thanks go to Professor Bart Smets of the Denmark Technical University, Copenhagen and Mark De Boer of the Rotterdam Zoo for kindly providing the strains used in this work.

Contents

Chapter 1 Introduction

1.1. Groundwater contamination by faecal matter

According to the WHO, an estimated one billion people lack access to an improved water supply and two million deaths per year are attributable to unsafe drinking water, sanitation and hygiene. In addition, many countries still report cholera to the WHO (WHO, 2004). Groundwater may be an important source of water for safe drinking and industrial water supplies, however, many water borne disease outbreaks are known to have been caused by the consumption of groundwater contaminated by pathogenic microorganisms (Goss et al., 1998; Macler and Merkle, 2000; Bhattacharjee et al., 2002; Close et al., 2006; Powell et al., 2003). Pathogenic microorganisms find their way into the sub-surface through the spreading of sewerage sludge on fields, leakage from waste disposal sites and landfills (Taylor et al., 2004), infiltration from cesspits, septic tank infiltration beds, and pit latrines (Foppen and Schijven, 2006), the application of human and animal excreta to land as crop manure (Gagliardi and Karns, 2002, Bolster et al., 2009, 2010) and pasturing of livestock, animal feeding operations (Gerba and Smith, 2005), thereby posing a threat to public health.

One of the explicit goals set by the United Nations and the international water community is environmental sustainability, with a target to halve the proportion of people without sustainable access to safe drinking water and basic sanitation by the year 2015. To achieve this Millenium Development Goal (MDG) of environmental sustainability, effective management and protection of water supply sources need to be practiced. One effective way of protecting groundwater sources from contamination by pathogenic microorganisms leaked into aquifer systems is by delineating well head protection areas around a drinking source. This strategy relies upon effective natural attenuation of sewage–derived microorganisms by soils (and rocks) over set back distances (Taylor et al., 2004). While natural processes may assist in reducing pollution, most biological contaminants can travel through soils and aquifers until they either enter someone's water well or are discharged into streams (Corapcioglu and Haridas, 1985). Although efforts have been made to understand the transport behaviour of bio-colloids in saturated porous media, still, much understanding is needed to improve prediction of interaction between bacterial cells and aquifer media.

1.2. Colloid filtration theory

To understand and predict microbial transport in the subsurface, studies are often performed in the laboratory and results obtained are applied to natural conditions. The retention and transport of microorganisms when passed through sand has commonly been determined with the classical colloid filtration theory (CFT; Yao et al., 1971; Tufenkji and Elimelech, 2004a). The theory is based on the assumption that colloids are retained at an invariable rate resulting in a log-linear reduction in deposition rate with increasing transport distance. The one dimensional (macroscopic) mass balance equation for mobile bacteria suspended in the aqueous phase excluding bacteria growth and decay is normally expressed as (Corapcioglu and Haridas, 1985.; Foppen et al., 2007a,b)

$$\frac{\partial C}{\partial t} = D\frac{\partial^2 C}{\partial x^2} - v\frac{\partial C}{\partial x} - \frac{\rho_{bulk}}{\theta}\frac{\partial S}{\partial t} \tag{1.1}$$

where C is the mass concentration of suspended bacteria in the aqueous phase (# of cells/ml), t is time (s), D is hydrodynamic dispersion coefficient (cm^2/s), v is velocity (cm/s), S is total retained bacteria concentration (#cells/gram sediment) ρ_{bulk} is the bulk density (g/ml), x is the distance traveled (cm) and θ is the volume occupied by the fluid per total volume medium (-). The first, second and third terms on the right hand side of equation (1.1) represent transport by hydrodynamic dispersion, advection and particle deposition, respectively. The retained bacteria fraction is given by a first-order kinetic term expression:

$$\frac{\partial S}{\partial t} = \frac{\theta}{\rho_{bulk}} k_a C \qquad (1.2)$$

where k_a (s^{-1}) is the attachment rate coefficient. For steady state conditions and negligible hydrodynamic dispersion and for continuous particle injection at concentration C_0 (at x=0) and time, t_0, the solution to equations (1.1) and (1.2) for a column initially free of particles is described by:

$$C(x) = C_0 \exp\left[-\frac{k_a}{v} x \right] \qquad (1.3)$$

and

$$S(x) = \frac{t_0 \theta k_a}{\rho_{bulk}} C(x) \qquad (1.4)$$

Equation (1.3) describes the colloid filtration theory (Yao et al., 1971, Iwasaki, 1937). Yao et al. (1971) described the attachment rate coefficient k_a (s^{-1}) as related to the single collector contact efficiency (η_0) (-) and the dimensionless sticking (collision) efficiency (α) by the following expression

$$k_a = \frac{3(1-\theta)\alpha\eta_0 v}{2d_c} \qquad (1.5)$$

where d_c is the mean grain (collector) diameter (cm). The collision efficiency (α) represents a percentage or fraction of colloids that successfully strike and stick to a collector surface.

The experimentally determined single collector removal efficiency (SCRE), η (-) and the attachment efficiency are also obtained under negligible hydrodynamic dispersion at steady state over total transport distance as

$$\eta = -\frac{2}{3} \frac{d_c}{(1-\theta)L} \ln\left(\frac{C}{C_0} \right) \qquad (1.6)$$

For a pulse injection, η can be quantified by replacing the relative breakthrough concentration in equation (1.6) with relative mass breakthrough (M_{eff} / M_{inf}) (Abudalo et al., 2005, Kretzschzmar et al., 1997) where M_{eff} and M_{inf} are the total number of cells in the effluent and influent, respectively.

The sticking efficiency is then obtained by

$$\alpha = \frac{\eta}{\eta_0} \qquad (1.7)$$

Tufenkji and Elimelech (2004a) developed a correlation equation to predict the single collector efficiency by summing the individual transport mechanisms to a collector surface as

$$\eta_0 = \eta_D + \eta_I + \eta_G \qquad (1.8)$$

where η_D (-), η_I (-) and η_G (-) are theoretical values for the SCCE when the sole transport mechanisms are diffusion, interception and sedimentation, respectively, and defined as

$$\eta_D = 2.4 A_S^{1/3} N_R^{-0.081} N_{Pe}^{-0.715} N_{vdW}^{0.052} \qquad (1.9a)$$

$$\eta_I = 0.55 A_S N_R^{1.675} N_A^{0.125} \qquad (1.9b)$$

$$\eta_G = 0.22 N_R^{-0.24} N_G^{1.11} N_{vdW}^{0.053} \qquad (1.9c)$$

where A_S (-) in equations (1.9a) and (1.9b) is a porosity dependent parameter defined as $A_S = 2(1-\gamma^5)/(2-3\gamma+3\gamma^5-2\gamma^6)$ and $\gamma = (1-\theta)^{1/3}$, $N_R = d_p/d_c$ interception number (-), d_p is the mean particle diameter (m). $N_G = 2d_p^2(\rho_p - \rho_f)g/9\varpi U$ is for sedimentation where g is the acceleration due to gravity (m/s²), ρ_p is the particle density (kg/m³) ρ_f is the fluid density (kg/m³), ϖ is the absolute fluid viscosity (Pa.s) and U is the fluid approach velocity (m/s). The Peclet number (-), $N_{Pe} = v\theta d_c/D_B$ for the sum of convection and diffusion. The van der Waals number expresses the ratio of van der Waals interaction energy to the particle's thermal energy $N_{vdW} = H/kT$ where H is the Hamaker constant (J). The expression $N_A = H/12\pi\varpi d_p^2 v\theta$ gives the attraction number (-) and expresses the combined influence of van der Waals attraction forces and fluid velocity on particle deposition rate due to interception. The description of the methodology for obtaining the correlation equation for each transport mechanism can be found in Tufenkji and Elimelech (2004a).

1.3. Deviation of bacteria transport from the colloid filtration theory

The classical colloid filtration theory is based on the assumption that the attachment of biocolloids to collector surfaces in saturated porous media is invariable and results in a log-linear reduction in fluid phase colloid concentration as travel distance increases. A characteristic of the classical theory is the use of the sticking efficiency, which is defined by the ratio of the number of particles that strike and stick to a collector to the number of particles striking a collector and is mainly determined by electro-chemical forces between the colloid and surface of the collector. Contrary to the classical theory, research results over the last two decades have indicated that the sticking efficiency of a biocolloid population is not a constant, and the variations have been attributed to variable cell surface properties of individual members of the population, resulting in differences in affinity for collector surfaces (Albinger et al., 1994; Baygents et al., 1998; Simoni et al., 1998; Li et al., 2004; Tufenkji and Elimelech, 2005a; Tong and Johnson, 2007; Foppen et al., 2007a,b). The variation of the deposition rate coefficient has been attributed to a number of reasons, including geochemical heterogeneity on collector grain surfaces (Johnson and Elimelech, 1996; Bolster et al., 2001; Loveland et al., 2003; Foppen et al., 2005), straining (Bradford et al., 2002 and 2003; Bradford and Bettahar, 2005; Foppen et al., 2007a,b), and heterogeneity of the colloid population due to variability in surface properties (Albinger et al.,1994; Baygents et al., 1998.; Simoni et al., 1998; Li et al., 2004; Tufenkji and Elimelech, 2005a,b; Tong and Johnson, 2007). The variability in bacteria surface properties has been attributed to variations in lipopolysaccharide (LPS) coating (Simoni et al.,1998), distribution of the interaction potential within the bio-colloid population (Li et al., 2004), variations in surface charge densities (Baygents et al.,1998; Tufenkji and Elimelech, 2004b), and differences in energy needed to overcome the energy barrier (Tufenkji and Elimelech, 2004b). Some group of workers (Redman et al., 2001a, b; Tufenkji et al. 2003) have demonstrated that a power-law best describes the distribution of sticking. Others found a log-normal distribution (Tufenkji et al., 2003; Tong and Johnson, 2007) or a dual distribution (Tufenkji and Elimelech, 2004b, 2005b; Foppen et al 2007a). The deviation of bacteria deposition patterns from the CFT has resulted in the inability to accurately predict transport distances in aquifers, with consequences of polluting drinking water sources (springs, boreholes and wells).

1.4. *Escherichia coli*

Escherichia coli (*E. coli*), a gram-negative, facultative non-spore forming, rod shaped bacterium is commonly used as indicator of faecal contamination of drinking water supplies, because *E. coli* is a consistent, predominantly facultative inhabitant of the human gastrointestinal tract. In addition, *E. coli* is easy to detect and quantify. Furthermore, the net negative surface charge and low inactivation rates of *E. coli* ensure that they may travel long distances in the subsurface and these characteristics make them a useful indicator for fecal contamination of groundwater (e.g. Foppen and Schijven, 2006). Due to the importance of *E. coli*, considerable attention has been given to understanding their transport and fate in saturated porous media (e.g. Foppen et al, 2007a,b, Schinner et al., 2010, Bolster et al., 2010). This thesis is looking into more detail at the transport of *E. coli* in saturated porous media.

1.5   Problem statement and objectives

Experimental results over the past decade-and-a-half have indicated that biocolloid interactions with saturated porous media vary within and among bacterial populations (Section 1.3). Results over the period revealed a transport distance dependent sticking efficiency, sticking efficiency distribution within bio-colloid populations and inter-strain attachment differences for different *E. coli* strains under given physico-chemical transport conditions. However, all studies, aimed at revealing sticking efficiency distributions, have been conducted for limited transport distances (centimeter to decimeter), and can therefore not be considered representative for longer transport distances, which are so important in microbial risk assessment of groundwater and therefore in quantifying the potential health impacts of pathogenic microorganisms traveling in aquifers. In addition, experiments that focussed on studying the effects of cell properties on their attachment to quartz grains have been conducted at short transport distances (<0.5 m) for limited number of strains (< 20 strains) (Becker et al., 2004, Bolster et al., 2010, Levy et al., 2007). Important questions still remaining are: How low can the sticking efficiency of fractions of cells within a population be? What are the effects of cell properties over long transport distances? And, how wide can the inter-strain attachment variability among substantial numbers (>20 strains) of different *E. coli* strains be? The objectives of this research were to:

- Study the inter-strain attachment variability for substantial (>20) numbers of *E. coli* strains, the effects of their phenotypic properties and genes encoding the outer membrane of *E. coli* cells on their attachment to quartz sand.

- Study the intra-strain attachment heterogeneities of *E. coli* strains, the distribution of sticking efficiencies over long transport distances (up to 25 m) and to measure low sticking efficiency values, that can be considered environmentally realistic. In addition, this study develops a methodology to measure the minimum sticking efficiency within *E. coli* sub-populations.

- Characterize the transport of *E. coli* strains isolated from springs, when considerable transport through different aquifers has already taken place. The underlying hypothesis was that transport by such a group of *E. coli* strains could possibly be characterized by a rather homogeneous set of strain characteristics and transport parameters.

1.6. Thesis outline

The first part of the thesis looks at the effects of *E. coli* cell properties on their transport in saturated porous media and consists of two chapters (2 and 3). In chapter 2, the effects of phenotypic characteristics (motility, hydrophobicity, outer surface potential and cell sphericity) of *E. coli* strains and an outer membrane protein (Antigen-43) on their attachment to quartz grains was studied over distances up to 5 m. In Chapter 3, inter-strain attachment differences amongst various *E. coli* strains (from soils and different parts of zoo animals) over 7 cm was studied. Furthermore, the effects of phenotypic characteristics and genes encoding 22 outer membrane structures of *E. coli* on attachment to quartz grains were investigated.

To measure environmentally realistic low sticking efficiencies and to develop a methodology to estimate the low values of bacteria attachment efficiency within bacterial sub-populations, Part II of the thesis focuses on the development of a methodology to determine the minimum sticking efficiency of *E. coli* strains, and it involves intra-strain attachment variations and distributions in *E. coli* cell affinity for quartz grain surfaces. In this second part, transport experiments were conducted over relatively long distances of up to 25 m in the laboratory. Chapters 4 and 5 form part II of the thesis. In Chapter 4, both the intra-strain and inter-strain heterogeneities were studied over a distance of 5 m in two solutions of different ionic strengths. The minimum sticking efficiency method developed in Chapter 4 was applied in Chapter 5 to study the transport of two environmental *E. coli* isolates.

In part III, methods to determine the cell characterization and transport of *E. coli* strains in Parts I and II were applied to study and measure cell properties and transport characteristics of *E. coli* strains isolated after they have been transported through aquifers. Chapters 6 and 7 form Part III. In chapter 6, inter-population and intra-population heterogeneities commonly observed with *E. coli* strains isolated from different sources were investigated using spring *E. coli* isolates, and chapter 7 focuses on bacteriological and physicochemical analyses of springs in the Lubigi catchment in Kampala, Uganda, and the transport of 40 *E. coli* strains isolated from the springs. The chapter also highlights on the variability in phenotypic characteristics of selected *E. coli* strains.

Summaries and conclusions of the research findings are presented in Chapter 8 which is the last part (Part IV) of this thesis.

PART I

EFFECTS OF *ESCHERICHIA COLI* PROPERTIES ON THEIR
TRANSPORT IN SATURATED POROUS MEDIA

Chapter 2 Effects of surface characteristics on the transport of multiple *Escherichia coli* isolates in large scale columns of quartz sand

This chapter is based on:
G. Lutterodt, M. Basnet, J.W.A. Foppen and S. Uhlenbrook (2009): *Effects of surface characteristics on the transport of multiple Escherichia coli isolates in large scale column of quartz sand. Water Research Vol. 43 p. 595-604.*

Abstract

Bacteria properties play an important role in the transport of bacteria in groundwater, but their role, especially for longer transport distances (> 0.5 m) has not been studied. Thereto, we studied the effects of cell surface hydrophobicity, outer surface potential, cell sphericity, motility, and Ag43 protein expression on the outer cell surface for a number of *E. coli* strains, obtained from the environment on their transport behavior in columns of saturated quartz sand of 5 m height in two solutions: demineralized water (DI) and artificial groundwater (AGW). In DI, sticking efficiencies ranged between 0.1-0.4 at the column inlet, and then decreased with transport distance to 0.02-0.2. In AGW, sticking efficiencies were on average 1 log unit higher than those in DI. Bacteria motility and Ag43 expression affected attachment with a (high) statistical significance. In contrast, hydrophobicity, outer surface potential and cell sphericity did not significantly correlate with sticking efficiency. However, for transport distances more than 0.33 m, the correlation between sticking efficiency, Ag43 expression, and motility became insignificant. We concluded that Ag43 and motility played an important role in E. coli attachment to quartz grain surfaces, and that the transport distance dependent sticking efficiency reductions were caused by motility and Ag43 expression variations within a population. The implication of our findings is that less motile bacteria with little or no Ag43 expression may travel longer distances once they enter groundwater environments. In future studies, the possible effect of bacteria surface structures, like fimbriae, pili and surface proteins on bacteria attachment need to be considered more systematically in order to arrive at more meaningful inter-population comparisons of the transport behavior of *E. coli* strains in aquifers.

2.1 Introduction

Groundwater systems globally provide 25 to 40% of the world's drinking water (Morris et al., 2003), and the importance of groundwater can often be attributed to the assumption that, in general, the resource is free of pathogenic microorganisms (Bhattacharjee et al., 2002). However, still in many cases, water borne disease outbreaks are caused by the consumption of groundwater contaminated by pathogenic microorganisms (Macler et al., 2000; Powell et al., 2003). Well-known sources of contamination are by leakage from septic tanks, unlined pit-latrines, improper waste disposals, manure, wastewater or sewage sludge (Foppen and Schijven, 2006). One of the explicit goals set by the United Nations and the international water community is environmental sustainability, with a target to halve the proportion of people without sustainable access to safe drinking water and basic sanitation by the year 2015. To achieve this Millenium Development Goal (MDG) of environmental sustainability, effective management and protection of water supply sources need to be practiced. Current strategies employed to protect groundwater sources from contamination rely upon effective natural attenuation of sewage–derived microorganisms by soils (and rocks) over set back distances (Taylor et al., 2004). While natural processes may assist in reducing pollution, most biological contaminants can travel through soils and aquifers until they either enter a water well or are discharged into streams (Corapcioglu and Haridas, 1985).

For a long time, the retention of microorganisms by passage through sand was determined with the classical colloid filtration theory (CFT; Yao et al., 1971; Schijven, 2001; Tufenkji and Elimelech, 2004a). The theory is based on the assumption that colloids are retained at an invariable rate, while deposition decreases log-linear with transport distance. However, recent research shows that the deposition rate coefficient is not a constant (Albinger et al., 1994; Baygents, 1998; Simoni et al., 1998; Li et al., 2004; Tufenkji and Elimelech, 2005a,b; Tong and Johnson, 2007; Foppen et al., 2007). The variation of the deposition rate coefficient has been attributed to a number of reasons, including geochemical heterogeneity on collector grain surfaces (Johnson and Elimelech, 1996; Bolster et al., 2001; Loveland et al., 2003; Foppen et al., 2005), straining (Bradford et al., 2002 and 2003; Bradford and Bettahar, 2005; Foppen et al., 2007a), and heterogeneity of the colloid population due to variability in surface properties (Albinger et al.,1994; Baygents et al., 1998.; Simoni et al., 1998; Li et al., 2004; Tufenkji and Elimelech, 2005a,b; Tong and Johnson, 2007; Foppen et al., 2007). The variability in bacteria surface properties has been attributed to variations in lipopolysaccharide (LPS) coating (Simoni et al.,1998), distribution of the interaction potential within the bio-colloid population (Li et al., 2004), variations in surface charge densities (Baygents et al.,1998; Tufenkji and Elimelech, 2004b), and differences in energy needed to overcome the energy barrier (Tufenkji and Elimelech, 2004b).

Three decades ago, a major outer membrane protein termed Antigen 43 (Ag43) was discovered in *Escherichia coli* (*E. coli*) (Das Gracas de Luna et al., 2008; Owen and Kaback, 1978). Outer membrane proteins serve a variety of functions essential to survival of Gram-negative bacteria. Many of these proteins have structural roles or are involved in transport, while others are important in pathogenesis and have roles in adhesion to host tissue or evasion of the host immune system (Nikaido, 2003). The outer membrane protein Ag43, encoded by the gene cluster *agn43*, was suggested as critical in determining the adhesive properties of *E. coli* (Henderson et al., 1997). Recently, Yang et al. (2004) isolated 280 *E. coli* strains from a soil, and they found that, under environmentally relevant growth conditions, the majority of *E. coli* isolates (88%) encoding Ag43 formed thick biofilms, while the majority of *E. coli* isolates not encoding Ag43 (75%) formed thin biofilms. Thus, Ag43 was involved in the attachment of *E. coli* cells.

Although studies have been conducted to determine the influence of LPS, bacteria growth stage and evolution of cell surface macro-molecules on cell adhesion (Walker et al., 2004 and 2005), the effect of bacteria properties on bacteria transport over longer travel distances (> 0.5 m) has not been systematically studied. In addition, the range of environmentally realistic sticking efficiencies, that determines bacteria travel distances in aquifers, are not known.

The objective of this chapter is to study the effects of a number of bacteria properties (Ag43 expression, motility, hydrophobicity, outer surface potential, and cell sphericity) of six *E. coli* strains on their transport behavior over environmentally realistic transport distances, up to 5 m.

2.2 Materials and methods

2.2.1 Extraction of manure and bacteria growth

We used six *Escherichia coli* (*E. coli*) strains (UCFL-71, UCFL-94, UCFL-131, UCFL-167, UCFL-263 and UCFL-348) obtained from a soil in a cattle grazing field(Yang et al., 2004). To mimic environmental conditions, *E. coli* isolates were grown in an extract of cow manure (Yang et al., 2006). Thereto, fresh cow manure was collected from a farm (cow ranch 'Ackersdijk', Delft, The Netherlands), and stored at -20 °C in batches of 50 g. Prior to each experiment, one batch of 50 g cow manure was defrosted and mixed with demineralized (DI) water at a 1:20 ratio (EPA – 1312 Leach Method). To facilitate extraction, the mixture was acidified to a pH of 5±0.05 with concentrated sulphuric acid and nitric acid at 60/40 weight percent mixture, and extraction was performed for 2 hours. The mixture was then centrifuged (IEC Centra GP 8- rotar 218/18cm) for 10 min at 4600 rpm (1185 g-force), and then at 9000 rpm for 10 min (816.5 g-force) (MSE high speed 18). The supernatant was sequentially filtered through a 0.45 μm and a 0.2 μm mesh size cellulose acetate membrane filter (47 mm diameter). *E. coli* isolates were activated from a glass test tube (pepton agar stock) and grown in Luria Bertoni (LB) broth (DifcoTM LB Broth, Miller) for 6 hours at 37 °C while shaking at 120 rpm on an orbital shaker. The inoculum was then diluted 10^5 fold in the cow manure extract and incubated, while shaking on the orbital shaker at 120 rpm, for 72 hours at 21 °C until a stationary growth phase was reached, resulting in a concentration of ~10^8 cells/ml.

2.2.2 Characterization of cell properties

We determined hydrophobicity, motility, outer surface potential of the *E. coli* cells, cell width and cell length, while data on Ag43 expression were obtained from Yang et al. (2006).They detected the presence of Ag43 using rabbit anti-Ag43α and FITC-labeled goat anti-rabbit serum and evaluated coagulation of cells using phase-contrast and epiflourescence micrographs. For a more detailed description we refer to Yang (2005).

Hydrophobicity was determined with the Microbial Adhesion To Hydrocarbons (MATH) method (Pembrey et al., 1999; Walker et al., 2005), where percentage partitioning of cells into dodecane was measured as cell hydrophobicity. Thereto, 4 mL of bacteria suspension of known optical density and 1mL of dodecane were vigorously mixed in a test tube for two minutes and left to stand for 15 minutes to allow phase separation. Then, the optical density of the aqueous phase was determined, and the percentage of cells partitioned into the hydrophobic substance was reported as percentage hydrophobicity. All optical densities were measured at an absorbance of 546 nm (Walker et al., 2005).

To determine the *outer surface potential* (OSP), a zeta-meter similar to the one made by Neihof (1969) was used. Movement of bacteria was visible on a video screen attached to a camera mounted on top of a light microscope (Olympus EHT) in phase contrast mode (Foppen et al., 2007b). Bacteria mobility values were obtained from measurements on at least 50 bacteria cells. Velocity measurements were used to calculate the OSP of the *E. coli* cells according to Ohshima's electrokinetic theory for soft particles (Ohshima, 1994; Dague et al., 2006). To determine the outer surface potential, we used the low potential approximation, given by De Kerchove and Elimelech (2005):

$$\psi_0 = \left(\mu - \frac{\rho_{fix}}{\eta\lambda^2}\right)\left(\frac{\eta}{\varepsilon_r\varepsilon_0}\right)\frac{\left(\frac{1}{\kappa} + \frac{1}{\lambda}\right)}{\left(\frac{1}{\kappa} + \frac{2}{\lambda}\right)} \tag{2.1}$$

whereby μ is the eletrophoretic mobility ($m^2V^{-1}s^{-1}$), ψ_0 is the OSP of an *E. coli* cell (V), ρ_{fix} is the fixed charge density of the polyelectrolyte layer at the *E. coli* surface (mol), η is the fluid dynamic viscosity (Pa s), $1/\lambda$ is the electrophoretic softness (m), ε_r is the dimensionless dielectric constant, ε_0 is the dielectric permittivity in a vacuum ($CV^{-1}m^{-1}$), and $1/k$ is the double layer thickness (m). For the calculations, we assumed the electrophoretic softness to be 0.74 nm^{-1} and the fixed charge density to be -150 mM. These values were taken from De Kerchove and Elimelech (2005) for an *E. coli* strain (D21g) with a soft layer of lipopolysaccharides in the outer membrane.

To determine *motility*, a cell suspension in cow manure extract was filtered through 0.2 µm cellulose acetate membrane filter, and by means of a sterile toothpick, cells were picked from the filter membrane and inoculated at the centre of petri-dishes containing 0.35% agar (Oxoid agar technical-agar no. 3), supplemented with manure extract. The plates were incubated at 23 °C for 48 hours after which growth and diameter of migration was measured as motility (Yang et al., 2006; Ulett et al., 2006).

To determine *width* and *length* of the cells, a light microscope (Olympus BX51) in phase contrast mode, with a camera (Olympus DP2) mounted on top and connected to a computer, was used to take images of cells. Averages of 50 images were imported into an image processing program (DP-Soft 2) and the average cell width and cell length were measured. The equivalent spherical diameter (ESD) was determined as the geometric mean of average length and width (Rijnaarts et al., 1993), while the cell sphericity was obtained from the ratio of average width to average length (Weiss et al., 1995).

2.2.3 Column experiments

To study the effects of the surface properties of the *E. coli* strains on their transport in the subsurface, column experiments were conducted in demineralised (DI) water and in artificial groundwater (AGW). The latter was prepared by dissolving 526 mg/L $CaCl_2.2H_2O$ and 184 mg/L $MgSO_4.7H_2O$ in DI water, and buffering it with 8.5 mg/L KH_2PO_4, 21.75 mg/L K_2HPO_4 and 17.7 mg/L Na_2HPO_4. The final pH-value ranged from 6.6 to 6.8 and the EC-value ranged from 1025 to 1054 μS/cm. The porous media comprised of 99.1% pure quartz sand (Kristall-quartz sand, Dorsilit, Germany) with sizes ranging from 180 to 500 μm, while the median of the grain size weight distribution was 356 μm. With this grain size, we excluded straining as a possible retention mechanism is our column: assuming a bacteria equivalent spherical diameter of 1.5 μm, the ratio of cell and grain diameter was 0.004. This was well below the ratio (0.007) for which straining was observed by Bradford et al. (2007) for carboxyl latex microspheres with a diameter of 1.1 mm suspended in solutions with ionic strengths up to 31mM (the ionic strength of the solutions we used was 4.7 mmol/L only). Total porosity was determined gravimetrically to be 0.40. Prior to the experiments, to remove impurities, the sand was rinsed sequentially with acetone, hexane and concentrated HCl, followed by repeated rinsing with DI water until the electrical conductivity was close to zero (Li et al., 2004).

The column consisted of a 5 m transparent acrylic glass (Perspex) tube with an inner diameter of 10 cm, and with seven sampling ports placed at 10-50 cm intervals along the tube. A stainless steel grid for supporting the sand was placed at the bottom of the tube. The column was gently filled with the clean quartz sand under saturated conditions, while the sides of the column were continuously tapped during filling, to avoid layering or trapping of air. The column was connected both at the funnel shaped effluent end and influent end with two Masterflex pumps (Console Drive Barnant Company Barrington Illinois, USA) via teflon tubes, and the pumps were adjusted to a mean fluid approach velocity of 1.16×10^{-4} m/s, coinciding with flushing the column with 1 pore volume (PV) per working day. We considered this fluid approach velocity fast enough to minimize the effects of die-off of *E. coli* on the measured bacteria concentrations at the various sampling ports. Prior to a column experiment, the column was flushed for two days with either DI or AGW to arrive at stable fluid conditions inside the column. Bacteria influent suspensions were prepared by washing and centrifuging at 3000 rpm for 10 minutes (90.7 g-force) three times in either DI or AGW, and then diluting 1000 times to arrive at bacteria cell concentrations of approximately 10^5 cells/mL. Experiments were conducted by applying a pulse of 0.15 PV (approximately 2.2L) of bacteria influent suspension to the column, followed by bacteria free DI or AGW. Samples were taken at 7 distances from the column inlet, and immediately plated (0.1 mL) in duplicate on Chromocult agar (Merck). Bacteria inactivation was assessed in all experiments by plating samples of the influent at an hourly frequency during the entire experiment. All plates were incubated at 37 °C for (at least) 18 hours.

After each experiment, to clean the sand in the column, and to prepare for the next experiment, a pulse of 5 L 1.9 M HCl followed by a pulse of 5 L 1.5 M NaOH was flushed through the column, followed by flushing with DI water until the electrical conductivity of the effluent was well below 3 μS/cm.

2.2.4 Computation of the sticking efficiency

Bacteria attachment to the sand was quantified by computing the sticking efficiency (α_L) (-) at the various sampling ports as (Kretzschmar et al., 1997, Abudalo et al., 2005)

$$\alpha_L = -\frac{2}{3}\frac{d_c}{(1-\theta)L\eta_0}\ln\left(\frac{M_{eff}}{M_{inf}}\right)$$
(2.2)

where d_c is the median of the grain size weight distribution (m), η_0 is the single collector contact efficiency (-), θ is the total porosity of the sand (-), L is the travel distance (m), M_{inf} is the total number of cells in the influent and M_{eff} is the total number of cells in the effluent (-) obtained as (Kretzschmar et al., 1997)

$$M_{eff} = q\int_0^t C(t)dt$$
(2.3)

where q is the volumetric flow rate (mL/min), C is the cell suspension (# cells/mL) and t is time (min). The TE correlation equation (Tufenkji and Elimelech, 2004a) was used to compute η_0. For this, we assumed that the bacteria density was 1055 kg/m^3, and the Hamaker constant was estimated to 6.5×10^{-21} J (Walker et al., 2004).

To assist in analyzing relationships between cell properties and sticking efficiency, we employed the strain averaged sticking efficiency ($\bar{\alpha}_{strain}$) defined as:

$$\bar{\alpha}_{strain} = \frac{\sum_{i=1}^{7}\alpha_{strain.L}}{7}$$
(2.4)

whereby subscript *strain* indicates the strain used and 7 refers to the number of sampling ports used.

2.2.5 Statistical analyses

The degree of association between cell properties and $\bar{\alpha}_{strain}$ for all strains and between cell properties and α_L for all traveled distances were determined with the parametric Pearson's correlation test. Correlations, determined with Pearson's correlation coefficient (r), were considered to be statistically significant, when $p \le 0.05$. All statistical analyses were conducted using SPSS 14 (SPSS Inc., 2005).

2.3 Results

2.3.1 Surface Properties

Average cell width for the strains ranged from 0.84 to 1.09 μm with a standard deviation (SD) of 0.09 μm, and the average length ranged from 1.76 to 2.42 μm (SD = 0.22 μm). The equivalent spherical diameter (Table 2.1) of the cells ranged from 1.33 to 1.53 μm (SD = 0.10 μm), while the cell sphericity ranged from 0.40-0.57 (SD = 0.07 μm). From this we concluded that the cell dimensions of all isolates used were within a relatively narrow range. The motility of the cells showed considerable variation (SD = 1.31 cm): while the diameter of the UCFL-167 colony on the soft agar plates was only 0.36 cm, the diameter of the UCFL-131 colony was 3.73 cm. Although not exactly similar, these variations had similar ranges as the variations measured by Yang (2005).

Table 2.1: Measured surface properties and Ag43 expression

Strain	ESD* (μm)	Sphericity (W/L)	Motility (cm)	Hydrophobicity (% partitioning)		Zeta potential (-mV)		Decay rate (hr^{-1})		Ag43 expression**
				DI	AGW	DI	AGW	DI	AGW	
UCFL-71	1.34	0.48	2.40	27.33	33.36	44.76	25.06	0.05	0.03	0
UCFL-94	1.45	0.54	0.57	22.12	44.80	54.74	17.89	0.01	0.00	1
UCFL-131	1.53	0.40	3.73	36.57	38.06	47.95	15.76	0.12	0.03	2
UCFL-167	1.30	0.42	0.36	81.01	55.96	31.79	23.91	0.02	0.01	1
UCFL-263	1.33	0.57	1.73	59.99	46.03	53.38	19.19	0.11	0.00	3
UCFL-348	1.53	0.51	2.77	35.37	53.45	49.36	24.30	0.07	0.01	3

*: ESD = Equivalent Spherical Diameter
**: 0: negative; 1: very few positive cells in a population; 2: positive; 3: positive, significant aggregates

The hydrophobicities varied from 22% (UCFL-94) to 81% (UCFL-167) in DI water (SD = 22 %) and from 33 to 55% in AGW (SD = 9 %). Four out of six strains (UCFL-71, UCFL-94, UCFL-131 and UCFL-348) were more hydrophobic in AGW than in DI water. The OSP ranged from -32 mV to -55 mV in DI water (SD = 8.3 mV) and from -14 mV to -22 mV in AGW (SD = 3.4 mV), so in all cases, the cells were negatively charged, while the less negative charge of the cells in AGW was explained by the compressed double layer due to the relatively high ionic strength compared to DI.

2.3.2 Breakthrough curves and sticking efficiencies

An example of one experiment with 7 normalized breakthrough curves, measured at the 7 sampling ports is shown in Fig. 2.1 for UCFL-263 in DI. The pulse with *E. coli* suspension was detected at every sampling port, while peak concentrations reduced with transported distance. Each of the other *E. coli* strains exhibited a similar trend in breakthrough curves (data not shown).

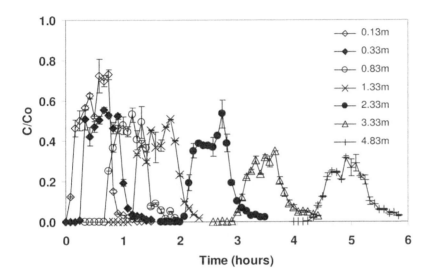

Figure 2.1:   Breakthrough curve of UCFL-263 in DI. Error bars indicate variation between two duplicate plate counts

In spite of the general trend, each strain exhibited a distinctly different transport behavior with respect to relative breakthrough at all 7 sampling ports. These differences in transport behavior also became apparent from the calculated sticking efficiencies for all *E. coli* strains at all sampling ports in both DI and AGW (Figs. 2.2 a and b). In both DI and AGW, a reduction of the sticking efficiency values was observed, with a comparatively stickier fraction of the cells being retained within 1.33 m from the column inlet, and a fraction exhibiting slow attachment (transport distance > 1.33 m). The total inter-strain sticking efficiency variations in both DI and AGW were around 1 log-unit (compare UCFL-94 and UCFL-348), while intra-strain sticking efficiency variation was around 0.5 log-unit in DI, and around 1 log-unit in AGW. UCFL-94 exhibited the lowest sticking efficiencies in both DI and in AGW, while UCFL-348 had highest sticking efficiencies. In AGW, UCFL-348 was completely retained within a travel distance of only 0.13 m.

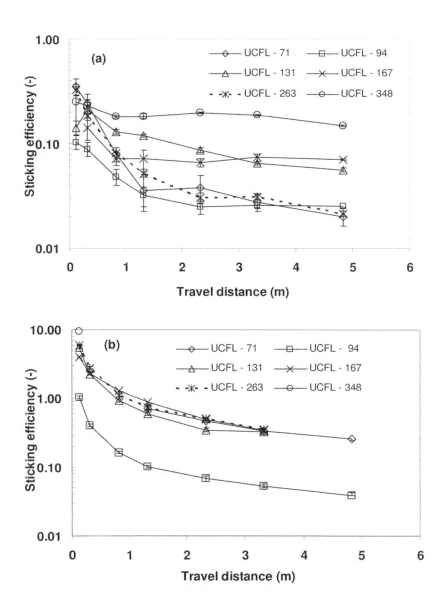

Fig 2.2:   *Sticking efficiency of the strains as a function of traveled distance in DI (a) and in AGW (b). Error bars indicate variation between two duplicate plate counts.*

2.3.3 Correlation between averaged sticking efficiencies and E. coli cell properties

In both DI and AGW there was no significant correlation between the strain averaged sticking efficiency, $\bar{\alpha}_{strain}$, and cell sphericity, outer surface potential or hydrophobicity (Tables 2.2 and 2.3).

Table 2.2:   Correlation matrix between the strain averaged sticking efficiency ($\bar{\alpha}_{strain}$) and cell surface properties for the experiments in DI. P values are in brackets and statistically significant ($p \le 0.05$) correlations are in bold.

	$\bar{\alpha}_{strain}$	Motility	Ag43	Hydrophobicity	Zeta potential
Motility	**0.724** **(0.052)**				
Ag43	**0.815** **(0.046)**	0.706 (0.092)			
Hydrophobicity	0.080 (0.440)	-0.261 (0.309)	0.064 (0.459)		
Zeta potential	-0.006 (0.496)	0.337 (0.257)	0.475 (0.209)	-0.634 (0.088)	
Sphericity	-0.110 (0.418)	-0.284 (0.293)	0.315 (0.303)	-0.25 (0.316)	0.663 (0.076)

Table 2. 3:   Correlation matrix between the strain averaged sticking efficiency ($\bar{\alpha}_{strain}$) and cell surface properties for the experiments in AGW. P values are in brackets.

	$\bar{\alpha}_{strain}$	Motility	Ag43	Hydrophobicity	Zeta potential
Motility	0.423 (0.202)				
Ag43	0.741 (0.076)	0.706 (0.092)			
Hydrophobicity	0.531 (0.139)	-0.515 (0.148)	-0.169 (0.393)		
Zeta potential	0.229 (0.332)	-0.218 (0.339)	0.033 (0.479)	0.294 (0.286)	
Sphericity	-0.080 (0.44)	-0.284 (0.293)	0.315 (0.303)	0.068 (0.449)	0.074 (0.444)

In order to avoid underestimating the low sticking efficiency values, when calculating the mean of all α_{strain} values per experiment, we also computed the average of the log-transformed values. However, the differences between the average sticking efficiency, $\bar{\alpha}_{strain}$, and the average of the log-transformed sticking efficiency were negligible, and we therefore did not consider log-transformed averaged sticking efficiencies. In DI, however, there was a high positive correlation between Ag43 expression and the strain averaged sticking efficiency ($r = 0.815$, $p = 0.046$) and also between motility and the strain averaged sticking efficiency ($r = 0.72$, $p = 0.052$). In AGW, strain averaged sticking efficiency showed a high positive correlation with Ag43-expression, but this correlation was not statistically significant ($r = 0.741$, $p = 0.076$). The correlation between strain averaged sticking efficiency and motility in AGW was low ($r = 0.42$, $p = 0.20$).

Correlations of pairs of all cell properties were generally low and statistically not significant. From these observations, we concluded that from the cell properties we studied, Ag43 expression was the most important property determining attachment of *E. coli*, while under certain conditions (the DI experiments) also motility played an important role.
We also determined the degree of correlation between the α_L-values of all strains and *E. coli* cell properties as a function of the transport distance. In both DI and AGW, the Pearson's correlation coefficient between sticking efficiency and Ag43-expression showed the highest positive r-values (Figs. 2.3 and 2.4)

In DI, r-values were not only positive, but also statistically significant for the first two sampling ports (at 0.13 m: $r = 0.87$, $p = 0.05$ and at 0.33 m: $r = 0.86$, $p = 0.02$). At larger transport distances, the Pearson's coefficient reduced, while the p-value increased, indicating less and less correlation between sticking efficiency and Ag43-expression as transport distance increased.

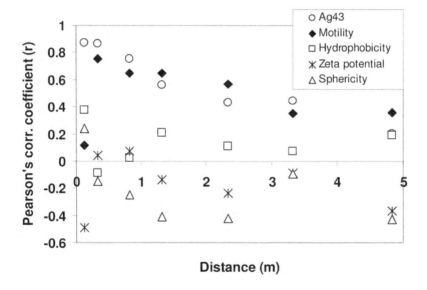

Fig.2.3:   *Pearson's correlation coefficient of the transport dependent sticking efficiency and E. coli surface property for the DI experiments, as a function of the traveled distance (m).*

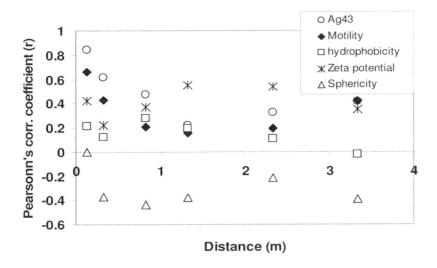

Fig. 2.4:   *Pearson's correlation coefficient of the transport dependent sticking efficiency and E. coli surface property for the AGW experiments, as a function of the traveled distance (m).*

In AGW a similar trend was observed: at 0.13 m, the correlation was high positive and statistically significant ($r = 0.846$, $p = 0.03$), and r reduced with increased transport distance.

In DI, the trends in correlations between α_l and motility were similar to Ag43-expression: at 0.13 m high positive and statistically significant ($r = 0.755$, $p = 0.035$) and the r-value decreased with increasing transport distance, while the p-value increased.

In AGW, however, r-values between α_l and motility were low and statistically not significant and also reduced with transport distance. Finally, at all travel distances in DI and AGW, the correlation between α_l and cell properties sphericity, hydrophobicity and outer surface potential were low and statistically insignificant.

2.4 Discussion

Our results showed that the Ag43 expression and motility significantly influenced bacteria attachment, while attachment was not significantly correlated with outer surface potential, sphericity and cell surface hydrophobicity. In addition, we also observed that the correlation between the Ag43 expression and sticking efficiency reduced with increased transport distance, and the same was true for the relation between motility and sticking efficiency.

2.4.1 Ag43 expression and motility

The importance of Ag43 expression in bacteria attachment to quartz grain surfaces was evident from the high positive r-value between Ag43 and $\bar{\alpha}_{strain}$ in both DI and AGW and from the high positive r-values between the sticking efficiency and Ag43 expression at the short travel distances (0.13 m and 0.33 m; in both solutions, DI and AGW) and reducing as travel distance increased. The Ag43 adhesin is known to establish autoaggregation of cells through Ag43-Ag43 interactions by a kind of intercellular handshake mechanism (Klemm et al, 2004; Hasman et al., 2000) resulting in the retardation of cell movement (Yang, 2005). In our case, such mechanism might have played a role, but, more importantly, we found that Ag43 also played a crucial role in the initial attachment of bacteria cells to the quartz grain surfaces. Ag43 is composed of two proteinaceous subunits, α^{43} and β^{43} (Henderson et al., 1997). Of these, α^{43} is surface expressed and is bound to the cell surface through an interaction with β^{43}, itself an integral, outer membrane protein. Although the mechanism involved in the contact between bacteria cells and grain surface is unknown, because of the predominantly negatively charged quartz surface, we surmise that in our case α^{43} was positively charged, and responsible for promoting α^{43} mediated favorable attachment between bacteria cell and quartz surface.

The reduction in r values between α_L and Ag43 expression with increasing transport distance can probably be attributed to intra-population heterogeneity. There is the possibility that not all cells within an Ag43 expressing strain indeed express the adhesin, which may lead to the differential transport and retention behavior (Figs. 2.3 and 2.4). We attribute the reduction in Pearson's correlation coefficient, r, with distance to preferential attachment of cells expressing the Ag43 adhesin. Also, Danese et al. (2000) reported that 43% of wild type *E. coli* cells grown in glucose minimal medium synthesize Ag43 and hypothesized the possibility of the remaining population expressing a partially redundant adhesin.

We found high positive r-values between motility and $\bar{\alpha}_{strain}$ in DI water and low positive r-values in AGW. In addition, the correlation between motility and α_L reduced with increasing transport distance. Motility might have increased the rate of diffusion of motile bacteria to the surfaces of the quartz grains. Motile bacteria can propel themselves to the collector surface, thereby overcoming repulsive forces (Pratt and Kolter, 1998), and thus increasing the rate of collisions. However, it remains difficult to explain the low r-value in DI at 0.13 m (Fig. 2.3). The high r values in DI water compared to the low r-values in AGW can be attributed to two factors:

(1)   In DI, non-motile cells were unable to overcome the repulsive electrostatic energy barrier between bacteria cell and quartz surface.

(2)   In AGW, the electrostatic repulsive barrier was low, and therefore both motile and non-motile cells could reach the surface of the quartz grains.

An increase in the rate of initial collisions between bacteria cells and collector surfaces, and a reduction in bacteria transport rates has also been attributed to flagella mediated motility (van Loosdrecht et al. 1989, Becker et al., 2004; Shemarova and Nesterov, 2005; de Kerchove and Elimelech, 2008). Their finding also explains our assertion that motility increases the probability of bacteria attachment in our experiments. In addition, Pratt and Kolter (1998) stated that the role of the flagellum which promotes bacteria motility in for example biofilm formation, is the promotion of initial contact. It should be noted that in our experiments we did not observe significant detachment of motile cells, evidenced by the lack of tailing of the breakthrough curves at distances between 0.13 m to 0.83 m (Fig. 2.1). As such, our observation contrasts with the view that motility might increase desorption by the liberation of bacteria from attachment bonds (McCallou et al., 1995).

2.4.2 Hydrophobicity, outer surface potential and cell sphericity

We observed very low r-values between hydrophobicity and the $\bar{\alpha}_{strain}$ in both DI and AGW. In addition, r-values along transport distance were low, and did not show any pattern. These observations are in contrast to those of van Loosdrecht et al. (1987a, b) and Jacobs et al. (2007). On the other hand, the observed non-dependence of cell hydrophobicity on attachment was in agreement with the findings of Gannon et al. (1991). It should be noted here that the MATH test is not straightforward for determining the hydrophobic character of bacteria. For example, Gaboriaud et al. (2006) demonstrated significant contributions of electrostatic interactions in such tests. However, the influence of pH and ionic strength on the percentage partition into dodecane is expected to be uniform across the strains, which allowed us to make a comparison of the various hydrophobicities measured in similar solutions.

Though cell shape has been shown to influence bacteria transport with preferential retention of elongated cells compared to more spherical cells (Weiss et al., 1995, Dong et al., 2002, Salerno et al. 2006), r-values between sphericity and $\bar{\alpha}_{strain}$ and also α_L were low-negative. Our results are consistent with Bolster et al. (2006), who also found that bacteria retention and cell sphericity were not correlated. We hypothesize that this non-dependence of attachment on sphericity in our case might be due to the narrow range of sphericities (0.40 to 0.57) of the E. coli strains we used. The outer surface potential (OSP) of the strains was not correlated with attachment in both DI and AGW. This is in contrast to the work of Sharma et al. (1985) and Foppen and Schijven (2006), who found a strong correlation between surface charge and attachment. Walker et al. (2005), however, observed differences in bacteria deposition rates though they recorded similar zeta-potential values. These workers stated that the deposition trend cannot be solely explained by electrostatic interaction due to the zeta-potential. In addition, de Kerchove and Elimelech (2005) demonstrated that the application of Ohshima's theory to their experimental data was inconsistent with known features of the E. coli cells they used. They attributed this inconsistency to chemical and physical inconsistencies associated with the ion-permeable polyelectrolyte layer at the cell surface. Such inhomogeneities were omitted in the development of Ohshima's theory. In our case, we concluded that the OSP is a lumped parameter that masked the actual interaction potential between individual cells and collectors. In studies whereby only one bacteria strain is used, variations of outer surface potential measurements can successfully give an indication of the processes involved in initial attachment, but in studies using more bacteria strains, due to the lumped character of the outer surface potential, the indicative value is completely lost.

2.4.3 Transport distance dependent sticking efficiency reduction

Sticking efficiencies decreased with increasing transport distance in both DI and AGW, by more than 1 log unit among the strains and less than 1 log within strains. In addition, in some cases, α_L values greater than 1 were found. The latter is not very strange (Shellenberger et al., 2002; Morrow et al. 2005; Paramonova et al. 2006), due to the presence of cell surface organelles like flagella and pili that extend beyond the cell surface. In such case, the aspect ratio used in the computation of the single-collector contact efficiency might be under estimated.

Our results are consistent with Simoni et al. (1998), Baygents et al. (1998), Redman et al. (2001a,b), Li et al. (2004) and Tufenkji and Elimelech (2004b, 2005b), who also reported a reduction of the sticking efficiency with transport distance. In some cases, those variable bacteria deposition rates were likely caused by variation in the LPS coating surrounding the bacteria cells, leading to heterogeneous interaction (Simoni et al., 1998), or by variability in surface charge densities within a bacteria population (Baygents et al.; 1998, van der Mei and Busscher, 2001, Tufenkji and Elimelech, 2004b). Our results indicated that these transport distance dependent sticking efficiency reductions were caused by the variable presence of motile cells and Ag43 expression: highly motile cells, expressing the Ag43 adhesin were removed faster than cells expressing only one of the two or neither one.

2.5 Conclusions

The effect of a number of bacteria properties (Ag43-expression motility, hydrophobicity, outer surface potential and sphericity) on attachment to quartz grain surfaces in columns of quartz sand up to 5 m were studied. Our results indicated that sticking efficiencies decreased with increasing transport distance in both DI and AGW, by more than 1 log unit among the strains and less than 1 log within strains. In addition, we found that Ag43 expression and motility significantly influenced bacteria attachment, while attachment was not correlated with outer surface potential, sphericity and cell surface hydrophobicity. Furthermore, we observed that the correlation between Ag43 expression and sticking efficiency reduced with increased transport distance, and the same was true for the relation between motility and sticking efficiency. Intra-population and inter-population heterogeneities exist within and among different *E. coli* strains, and the prediction of transport distances based on experimental results with a single strain cannot be simply extrapolated. The implication of our findings is that less motile bacteria with little or no antigen expression may travel longer distances once they enter groundwater environments. In future studies, the possible effect of bacteria surface structures, like fimbriae, pili and surface proteins on bacteria attachment need to be considered more systematically in order to arrive at more meaningful inter-population comparisons of the transport behavior of *E. coli* strains in aquifers.

Chapter 3 Towards understanding inter-strain attachment variations of *Escherichia coli* during transport in saturated quartz sand

This chapter is based on:
J. W. Foppen, G. Lutterodt, W. Rölling, and S. Uhlenbrook (2010): Towards understanding inter-strain attachment variations of Escherichia coli during transport in saturated quartz sand. Water Research Vol. 44 p. 1202-1212.

Abstract

Although *E. coli* is an indicator of fecal contamination in aquifers, limited research has been devoted to understanding the biological processes involved in the initial attachment of *E. coli* transported in abiotic porous media. The roles of the various surface structures of *E. coli*, like lipopolysaccharides (LPS), auto-transporter proteins, and fimbriae are unknown. The objective of this research was to establish the effects of variations in surface characteristics of the outer membrane of *E. coli* on the attachment efficiency of 54 *E. coli* strains upon transport in saturated quartz sand under identical flow conditions. We used column experiments to assess retention of the *E. coli* strains, and we determined sphericity, motility, zeta-potential, and aggregation of all strains. LPS composition was determined based on known serotypes, and the presence/absence of 22 genes encoding surface characteristics was determined with qualitative PCR. The results indicated that under identical flow conditions, there was a variation of two orders of magnitude in the maximum breakthrough concentrations of the 54 *E. coli* strains. Of all factors we investigated, no single factor was able to explain attachment efficiency variations statistically significantly. However, low attachment efficiencies were associated ($p = 0.13$) with LPS containing saccharides with phosphate and/or carboxyl groups. These saccharide groups are acidic and likely charged with a negative O-atom, which reduced attachment to the negatively charged quartz surface. In addition, of the 22 genes tested, *Afa* was most associated ($p = 0.21$) with attachment efficiency. The work presented here bridges knowledge on colloid transport and molecular microbiology, and tries to offer a more holistic view on the attachment of planktonic *E. coli* bacteria to (abiotic) quartz grain surfaces. Future research should evaluate the use of microbiological techniques in order to be able to map the unique or grouped characteristics of *E. coli* in aquifers, and to assess the usefulness of *E. coli* as a fecal indicator in aquifers.

3.1 Introduction

Being an important member of the normal intestinal micro-flora of humans and other mammals, *Escherichia coli* has been widely used as an indicator micro-organism of fecal pollution (Medema et al., 2003). But *E. coli* is more than just a harmless intestinal inhabitant; it can also be a highly versatile, and frequently deadly, pathogen (Kaper et al., 2004). The most notorious *E. coli* serotype is O157:H7 (Stenutz et al., 2006), which has been the cause of several large outbreaks of disease in North America, Europe and Japan, related to drinking water (Hrudley et al., 2003) and food (Grimm et al., 1995; Kaper, 1998; Ozeki et al., 2003; Ezawa et al., 2004). This *E. coli* O157:H7 is a so called entero-hemorrhagic *E. coli* (EHEC), causing bloody diarrhea (haemorrhagic colitis), non-bloody diarrhea and hemolytic uremic syndrome (Boyce et al., 1995). Upon transport in aquifers, a variety of processes can have an impact on the interactions of the traveling organism with aquifer material, resulting in different apparent travel velocities and concentration changes along a flow line. One of the more important processes is the interaction of the surface of the organism with the aquifer grain surface (Foppen and Schijven, 2006).

The *E. coli* surface contains several different structures, especially lipopolysaccharides, autotransporter proteins, flagella, fimbriae, adhesins, curli, and porins (Van Houdt and Michiels, 2005). Lipopolysaccharides (LPS), also known as endotoxins, are anchored in the outer membrane of *E. coli*. They consist of a common lipid A, a core region, and an O-antigen polysaccharide, which is specific for each serogroup. LPS occupies 75% of the surface of the bacterium, and *E. coli* is estimated to have 10^6 molecules per cell (Caroff and Karibian, 2003). More than 180 different O-serotypes have been described, but the exact saccharide composition of more than 50% of all known *E. coli* serotypes is still unknown (Stenutz et al., 2006).

Autotransporter proteins are secretory proteins, consisting of various units, causing large polyproteins with an aminoterminal domain extending from the cell surface into the environment (Henderson et al., 1998). Adhesive phenotypes have been attributed to a subfamily of *E. coli* autotransporters, including Ag43, AIDA and TibA. Antigen 43 (Ag43) is a prominent surface protein of *E. coli* (around 50000 copies per cell). This autotransporter protein is a self-recognizing adhesin, which contains both receptor recognition and receptor target, and protrudes approximately 10 nm beyond the outer membrane (Van Houdt and Michiels, 2005). Flagella are made up of the protein flagellin.

The role of flagellar filaments, motility and chemotaxis in biofilm formation has been well established (Van Houdt and Michiels, 2005). Also, fimbriae have been associated with host tissue adhesion of important pathogenic *E. coli* strains. An overview of the most important fimbriae and adhesins, mostly related to virulence factors of *E. coli* is given in Table 3.1. The most common adhesins found in both commensal and pathogenic *E. coli* isolates are Type 1 fimbriae, which are 7-nm wide, approximately 1-µm long rod-shaped adhesive surface organelles (Van Houdt and Michiels, 2005). A type I fimbriated cell can have up to 500 fimbriae, consisting of up to 5 million proteins, representing about 8% of the total cellular protein (Schembri and Klemm, 2001). Their importance in biofilm formation has been well studied (Van Houdt and Michiels, 2005). *E. coli* strains often produce other fimbriae (Table 3.1) which are classified on the basis of their adhesive, antigenic or physical properties or on the basis of similarities in the primary amino acid sequence of their major protein subunits. The involvement of these other types of fimbriae in biofilm formation has been studied, but to a much lesser extent (Van Houdt and Michiels, 2005).

With regard to the other surface structures referred to in Table 3.1, *ompC* and *slp* form outer membrane proteins during the initial stages of biofilm formation (Sauer, 2003). Recently, Tabe Eko Niba et al. (2007) demonstrated that mutants defective of *surA* were highly incapable of forming biofilms, possibly due to the lack of initially attaching to abiotic surfaces.

A large amount of research has been devoted to understanding *E. coli* attachment to biotic surfaces. In contrast, limited research has been devoted to understanding the mechanisms involved in the initial attachment of *E. coli* transported in abiotic porous media, and the roles these various surface structures described above, may play. Knowledge about these processes is vital when making assessments of the suitability of using *E. coli* as an indicator organism for specific pathogenic microorganisms in groundwater, for modeling and understanding the movement of *E. coli* in the subsurface, or, more in general, for the application and injection of bacteria in bioremediation studies. A good example of a more recent study was carried out by Walker et al. (2004), who looked at the influence of LPS composition on cell adhesion. These authors conclude that a complex combination of cell surface charge heterogeneity and LPS composition is in control of the adhesive characteristics of *E. coli* K12. Lutterodt et al. (2009a) studied the transport of 6 *E. coli* strains in 5 m columns of saturated quartz sand. These authors conclude that Ag43 and motility play an important role in *E. coli* attachment to quartz grain surfaces. In addition, they found that attachment efficiencies reduced with transport distance, and these reductions were possibly related to motility variations and Ag43 expression variations within an *E. coli* population. Finally, Bolster et al. (2009) show that there is a large diversity in cell properties and transport behavior for the 12 different *E. coli* isolates they used. With the parameters they used to characterize the *E. coli* surface (electrophoretic mobility, cell size and shape, hydrophobicity, charge density, and extracellular polymeric substance composition), they were not able to explain *E. coli* attachment variations.

The objective of this research was to establish the effects of variations in surface characteristics of the outer membrane of *E. coli* on the attachment efficiency of 54 *E. coli* strains upon transport in saturated quartz sand under identical flow conditions. In addition, we attempted to determine which of these genes encoding those structures at the *E. coli* surface were likely related to the initial attachment of the strains we used.

Table 3.1: Genes encoding surface structures involved in initial attachment of Escherichia coli including primers and DNA sequences

Gene name	Description / function	Primer name	DNA sequence (5' -> 3')	Amplified product (bp)	Source
	Autotransporter proteins:				
flu (Ag43)	Adhesin involved in diffuse adherence, consisting of an Ag43α and an Ag43β unit	Ag43-550-F	GTACKRCCAACGGAATGACC	515	This study
		Ag43-550-R	ATCCAGYGCGTRRCCATGTA		
		Ag43-2700-F	GACTATGACCGGATTNTGGC	719	This study
		Ag43-2700-R	GGCTGTACCCACCAGTTCAC		
aidA	Adhesin involved in diffuse adherence (AIDA-1)	UN21	TGCAAACATTAAGGGCTCG	450	Chapman et al., 2006
		UN22	CCGGAAACATTGACCATACC		
aidA	Adhesin involved in diffuse adherence (AIDA-C)	UN23	CAGTTTATCAATCAGCTCGGG	543	Chapman et al., 2006
		UN24	CCACCGTTCCGTTATCCTC		
aah	Autotransporter adhesin heptosyltransferase encoding AAH protein which modifies AIDA-I adhesin	UN19	CTGGGTGAGCATTATTGCTTGG	370	Chapman et al., 2006
		UN20	TTTGCTTGTGCGGTAGACTCG		
	Pili/fimbriae:				
afa/draBC	Central region of Dr antigen-specific fimbrial and afimbrial adhesin operons (e.g., AFA, Dr, and F1845)	Afa-F	GGCAGAGGGCGCGCAACAGGC	559	Chapman et al., 2006
		Afa-R	CCCGTAACGCGCCAGCATCTC		
bfpA	Type IV bundle-forming pili	bfpA-F	AATGGTGCTTGCGCTTCGTGC	326	Chapman et al., 2006
		bfpA-R	GCCGCTTTATCCAACCTGGTA		
bmaE	M-agglutinin subunit	bmaE-F	ATGGCGCTAACTTGCCATGCTG	507	Chapman et al., 2006
		bmaE-R	AGGGGGACATATAGGCCCCTTC		
facG	F4 fimbrial adhesin	facG-F	GGTGATTTCAATGGTTCG	764	Chapman et al., 2006
		facG-R	AATTGCTACGTTCAGCGGAGCG		
fanC	F5 fimbrial adhesin	fanC-F	TGGGACTACCAATGCTTCTG	450	Chapman et al., 2006
		fanC-R	TATCCACCATTAGACGGAGC		
fedA	F18 fimbrial adhesin	fedA-F	GTGAAAAGACTAGTTTATTTC	510	Chapman et al., 2006
		fedA-R	CTTGTAAGTAACCGCGTAAGC		
fimH	D-Mannose-specific adhesin, type 1 fimbriae	FimH-F	TGCCAGAACGGCGATAAGCCGTGG	508	Chapman et al., 2006
		FimH-R	GCAGTCACCTGCCCTCCGGTA		
focG	Pilus tip molecule, F1C fimbriae (sialic acid specific)	focG-F	CAGCACGGCAGTGGGAIACGA	360	Chapman et al., 2006
		focG-R	GAATGTCGCCTGCCCATTGCT		
iha	Novel nonhemagglutinin adhesin (from O157:H7 and CFT073)	iha-F	CTGGCGGAGGCTCTGAGATCA	827	Chapman et al., 2006
		iha-R	TCCTTAAGCTCCCGCGGCTGA		
nfaE	Nonfimbrial adhesin I assembly and transport	nfaE-F	GCTTACTGIATTCTGGGATGGA	559	Chapman et al., 2006
		nfaE-R	CCGTGGCGAGICAATAIGCCA		
paa	Porcine A/E-associated gene	M155-F1	ATGAGGAAACATAATGGCAGG	350	Chapman et al., 2006
		M155-R1	TCTGGTCAGGTGCGTCAATAC		
saa	STEC autoagglutinating adhesin	Saa-F	CGTGATGAACAGGGCTATTGC	119	Chapman et al., 2006
		Saa-R	ATGGACATTGCCTGTGGGCAAC		
sfa/focDE	Central region of sfa (S fimbriae) and foc (F1C fimbriae) operons	sfaI-F	CTCCGGAGAACTGGGTGCATCTTAC	410	Chapman et al., 2006
		sfaI-R	CGGAGGAGTAATTTACAAACCTGGCA		
sfaS	Pilus tip adhesin, S fimbriae (sialic acid specific)	sfaS-F	GTGGATACGACGATTACTGTG	240	Chapman et al., 2006
		sfaS-R	CCGCCAGCATTCCCTGTATTC		
	Other:				
ompC	Forms an outer membrane protein which is involved in biofilm formation	ompC-F	CTACATGCGTCTTGGCTTCA	633	This study
		ompC-R	GTTGCGTTRTARGTCTGGGT		
slp	Forms an outer membrane lipoprotein induced after carbon starvation (initial steps in biofilm formation)	slp-F	GGCATACTGGGCAGGTACGTT	439	This study
		slp-R	GCATAATCACCTGCWGGCGTT		
surA	Participates in the assembly of outer membrane proteins	surA-F	TTGCTAACATTGCGAAACAG	651	This study
		surA-R	CACTCTTGATATCRGCAGCA		

3.2 Materials and Methods

3.2.1 Bacteria strains.

We used 54 *E. coli* strains. Of those, 48 strains were obtained from the Zoo in Rotterdam, The Netherlands; six other strains we used in previous research (Lutterodt et al., 2009; UCFL strains in Table 3. 2). The Zoo strains were isolated during the last 5-10 years from various animals living in the Zoo, and from various sources (liver, kidney, feces, lungs, etc.). All Zoo strains were previously serotyped by the National Institute of Public Health and the Environment of the Netherlands. Strains were grown in 50mL of nutrient broth (Oxoid CM001) for 24 h at 37 °C. Bacteria were washed and centrifuged (14000xg) three times in Artificial Ground Water (AGW). AGW was prepared by dissolving 526 mg/L $CaCl_2.2H_2O$ and 184 mg/L $MgSO_4.7H_2O$, and buffering with 8.5 mg/L KH_2PO_4, 21.75 mg/L K_2HPO_4 and 17.7 mg/L Na_2HPO_4. The final pH-value ranged from 6.6 to 6.8 and the EC-value ranged from 1025 to 1054 µS/cm. In this way, we hoped to create an environment inside the column with a low repulsive double layer energy barrier, in order to enhance attachment of cells.

3.2.2 Porous media

The porous media comprised of 99.1% pure quartz sand (Kristall-quartz sand, Dorsilit, Germany) with sizes ranging from 180 to 500 µm, while the median of the grain size weight distribution was 356 µm. With this grain size, we excluded straining as a possible retention mechanism in our column: assuming a bacteria equivalent spherical diameter of 1.5 µm, the ratio of colloid and grain diameter was 0.004, which was well below the ratio (0.007) for which straining was observed by Bradford et al. (2007) for carboxyl latex microspheres with a diameter of 1.1 mm suspended in solutions with ionic strengths up to 31mM (the ionic strength of the solutions we used was 4.7 mmol/L only). Total porosity was determined gravimetrically to be 0.40. Prior to the experiments, to remove impurities, the sand was rinsed sequentially with acetone, hexane and concentrated HCl, followed by repeated rinsing with de-mineralized water until the electrical conductivity was close to zero (Li et al., 2004).

3.2.3 Column experiment

Column experiments were conducted in borosilicate glass columns with an inner diameter of 2.5 cm (Omnifit, Cambridge, U.K.) with polyethylene frits (25 µm pore diameter) and one adjustable endpiece. The column was packed wet with the quartz sand with vibration to minimize any layering or air entrapment. Column sediment length was 7 cm. All column experiments were conducted in artificial groundwater (AGW) at a velocity of 0.25 PV per minute (fluid approach velocity = 10^{-4} m/s). To eliminate *E. coli* retention variations resulting from variations in packing of the sand, all experiments were carried out in one column, packed at the start of the entire set of column experiments. Prior to each experiment, and in order to remove retained cells of the previous experiment, the column was rinsed with 1 PV of 1.9 M HCl, immediately followed by a pulse of 1.5 M NaOH to restore pH. Then, the column was equilibrated 50-60 pore volumes with AGW in order to restore pH and EC. Usually, a suspension of *E. coli* with a concentration of ~ 10^8 cells/mL was flushed through the column for 4 minutes (approximately equal to one pore volume) followed by a flush of *E. coli*-free AGW. The *E. coli* concentration was determined using optical density measurements (at 410 nm) with a 1 cm flow-trough glass cuvette and a spectrophotometer (Cecil 1021, Cecil Instruments Inc., Cambridge, England). Absolute cell numbers were deduced after calibration with plate counts on Chromocult™ agar (Merck, Whitehouse Station, NJ). To check whether the flush with HCl followed by NaOH had indeed removed all bacterial cells, at the beginning of each experiment, effluent samples were plated in triplicate. All plates of all experiments were negative, indicating that, after the previous experiment, all viable bacterial cells had indeed been removed from the column. To check for consistency of the methodology, at the end of the entire set of experiments, 14 experiments with varying breakthrough were repeated in random order. Breakthrough curves of all duplicate column experiments were nearly identical, and the two group mean attachment efficiencies were identical. From this, we concluded that the methodology we used was consistent and yielded results that could be reproduced. In the absence of straining, we assumed that retention was predominantly characterized by attachment of *E. coli* cells to the quartz grain surfaces in the column. Bacteria attachment to the sand was quantified by computing the attachment efficiency (α) as (Lutterodt et al., 2009a; Kretzschmar et al., 1997; Abudalo et al., 2005)

$$\alpha = -\frac{2}{3} \frac{d_c}{(1-\theta)L\eta_0} \ln\left(\frac{M_{eff}}{M_{inf}}\right) \tag{3.1}$$

where d_c is the minimum of the grain size weight distribution (m), η_0 is the single collector contact efficiency (-) , θ is the total porosity of the sand (-), L is the travel distance (m), M_{inf} is the total number of cells in the influent and M_{eff} is the total number of cells in the effluent (-) obtained as (Kretzschzmar et al., 1997)

$$M_{eff} = q\int_0^t C(t)dt \tag{3.2}$$

where q is the volumetric flow rate (mL/min), C is the cell suspension (# cells/mL) and t is time (min). The TE correlation equation (Tufenkji and Elimelech, 2004a) was used to compute η_0. For this, we assumed that the bacteria density was 1055 kg/m^3, and the Hamaker constant was estimated to 6.5×10^{-21} J (Walker et al., 2004).

3.2.4 Cell characterization

To determine *width* and *length* of the cells, a light microscope (Olympus BX51) in phase contrast mode, with a camera (Olympus DP2) mounted on top and connected to a computer, was used to take images of cells. Averages of 50 images were imported into an image processing program (DP-Soft 2) and the average cell width and cell length were measured. The equivalent spherical diameter was determined as the geometric mean of average length and width (Rijnaarts et al., 1993), while the cell sphericity was obtained from the ratio of average width to average length (Weiss et al., 1995).

To determine *motility*, a 2 mL fresh culture was centrifuged (14000 xg) and washed three times in AGW, and by means of a sterile toothpick, cells were picked from the remaining pellet in the test tube and inoculated at the centre of petri-dishes containing 0.35% Chromocult agar. The plates were incubated at 37 °C for 24 hours after which growth and diameter of migration was measured as motility (Ulett et al., 2006).

To determine the *zeta potential*, a zeta-meter similar to the one made by Neihof (1969) was used. Movement of bacteria was visible on a video screen attached to a camera mounted on top of a light microscope (Olympus EHT) in phase contrast mode as reported by Foppen et al.(2007). Bacteria mobility values were obtained from measurements on at least 50 bacteria cells. Velocity measurements were used to calculate the zeta potential with the Smoluchowksi equation.

To determine *auto-aggregation*, 15 ml of freshly grown bacteria were centrifuged (14000 xg) and washed three times in AGW, and, then, allowed to stand for 180 minutes at a temperature of 4 °C. A sample of 1 mL 1cm below the surface of the suspension was obtained, immediately and 180 minutes after washing. The optical density of the samples was measured at 254 nm, and the auto-aggregation was determined as the ratio of the final over the initial optical density (in %).

3.2.5 Serotypes and lipopolysaccharide structure

Most of the information on polysaccharide structure of the various serotypes can be found in a database on the Internet (www.casper.organ.su.se/ecodab). The database is described in detail by Stenutz et al. (2006). Since 2006, the polysaccharide structure of a limited number of additional *E. coli* serotypes was elucidated, and included in this research.

3.2.6 Detection of genes encoding factors related to E. coli surface structures

The polymerase chain reaction (PCR) is a powerful technique to amplify a single or few copies of a piece of DNA (here, genes possible involved in attachment) for several orders of magnitude, generating millions or more copies of a particular DNA sequence. The method relies on thermal cycling, consisting of cycles of repeated heating and cooling of the reaction for DNA melting and enzymatic replication of the DNA. Prior to PCR, bacterial cultures were pre-grown in nutrient broth (OXOID CM 001) for 24 h at 37°C. DNA of *E. coli* cells was isolated with a FastDNA Spin Kit (QBiogene), involving a mechanical 'bead-beating' procedure.

Qualitative PCR was carried out for the ten strains with highest attachment efficiencies and for the ten strains with lowest attachment efficiencies in 48-well plates in a 25 μL volume containing 1 μl of isolated DNA, 1 μl of both forward primer (10 μM) and reverse primer (10 μM), 12 μl FideliTaq PCR Master Mix 2× (USB Cooperation, Ohio, USA), and 10 μl MilliQ water. PCR conditions consisted of an initial denaturation step (4 min 94 °C), 25 cycles of denaturation (at 92 °C for 1 min), annealing (dependent on primer) and extension (72 °C for 1 minute), and a final elongation cycle (5 min at 72 °C). Primers used are given in Table 3.1. Primers prepared in this study (Table 3.1) were determined with PRIMERBLAST by using the *E. coli* genes as query sequences (www.ncbi.nlm.nih.gov). PCR mixtures were electrophoresed on a 1.2% agarose gel stained with ethidium bromide (0.08 μg/ml) for 30 minutes at 80-100 V. Results were made visible under UV light (302 nm) with a UV transilluminator.

3.2.7 Statistical analyses

Correlation between attachment efficiency and sphericity, motility, and zeta potential were determined with Pearson correlation. Correlation between attachment efficiency and cell-aggregation was determined using the Spearman rank correlation test. Correlation between attachment efficiencies, saccharides and genes were determined using Fisher's exact test, and correlation between cell aggregation and the presence of the *flu* gene was determined using the Kruskal-Wallis test. Most of the tests were carried out in Systat (SPSS Inc. Chicago, Il), except for Fisher's test (www.socr.ucla.edu).

3.3 Results

3.3.1 Breakthrough curves

In general, breakthrough was rapid and started to increase within one pore volume (data not shown). Not all curves started to breakthrough at exactly the same (dimensionless) moment. This was possibly due to an equilibrium sorption component, which caused minor horizontal translations of the breakthrough curves. In all cases, after the bacteria were flushed, relative concentrations rapidly fell to values below the detection limit of the spectrophotometer.

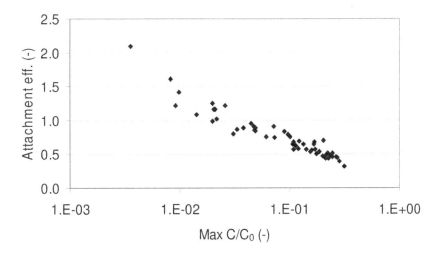

Fig. 3.1:   *Relation between maximum recorded breakthrough and attachment efficiency for the 54 E. coli strains used in this study.*

The maximum recorded relative breakthrough was 0.32 (UCFL-94; Table 3.2), and minimum breakthrough was 0.0036 (strain 2049; Table 3.2). From this we concluded that under these completely identical flow conditions in our 7 cm column, variations in the characteristics of the *E. coli* strains had caused relative breakthrough variations of two log units (Fig. 3.1). In general, calculated attachment efficiencies (Table 3.2) ranged from 0.3 to 1. However, a number of calculated attachment efficiencies were above unity, which is theoretically impossible.

Table 3.2: *Escherichia coli strains used in this study, including serotypes, and experimentally determined parameters, ordered from low to high attachment efficiency*

Strain	Serotype	Max C/C$_0$ (-)	Attachment eff. (-)	Width (µm)	Length (µm)	Sphericity (-)	Motility (cm)	Zeta pot. (-mV)	Aggregation (%)
UCFL-94	?	0.318	0.32	1.91	3.73	0.51	0.23	27.0	4.46
UCFL-131	?	0.286	0.40	1.90	2.84	0.67	5.73	27.7	3.25
UCFL-71	?	0.224	0.44	1.71	3.19	0.54	0.13	20.1	2.40
1923	O139	0.271	0.44	1.99	3.39	0.59	2.67	24.0	0.24
2043	O29	0.263	0.45	1.58	2.98	0.53	3.93	25.3	0.15
2007	O45	0.211	0.46	1.59	2.78	0.57	3.73	18.4	0.96
2160	O124	0.246	0.46	1.81	3.45	0.52	3.13	22.9	2.26
1991	O1	0.236	0.47	1.89	3.07	0.62	6.10	21.6	3.44
2621	O5	0.200	0.49	1.61	2.65	0.61	2.67	20.3	4.59
1924	O110	0.245	0.50	1.76	2.66	0.66	6.27	28.4	0.13
UCFL-263	?	0.217	0.51	1.64	2.61	0.63	4.73	24.3	4.81
UCFL-167	?	0.220	0.51	1.76	3.45	0.51	2.60	32.1	6.16
2216	O58	0.222	0.51	1.55	2.77	0.56	0.97	19.5	0.16
1876	O7	0.176	0.52	1.65	2.86	0.58	4.43	25.4	2.07
2312	O83	0.186	0.54	1.78	2.92	0.61	2.27	17.9	1.83
1927	O16	0.158	0.54	2.00	3.44	0.58	1.67	19.7	1.63
2266	O76	0.154	0.56	1.81	2.78	0.65	7.93	26.1	2.10
2000	O20	0.171	0.57	1.66	3.68	0.45	0.80	19.8	0.59
2264	O71	0.142	0.57	2.07	3.13	0.66	2.80	19.2	0.14
2257	O96	0.108	0.57	1.57	2.71	0.58	3.00	21.1	3.43
2269	O148	0.120	0.58	1.99	2.92	0.68	5.20	23.0	2.31
2260	O49	0.166	0.62	1.89	3.55	0.53	2.27	21.9	0.90
1366	O23	0.114	0.64	1.74	3.05	0.57	5.30	28.5	1.55
1870	O166	0.204	0.64	1.80	3.56	0.51	4.47	24.5	6.29
1875	O2	0.133	0.65	1.45	2.31	0.63	0.80	30.6	1.66
2059	O59	0.167	0.68	2.02	3.22	0.63	4.37	20.6	4.01
2602	O159	0.107	0.68	1.48	2.39	0.62	3.90	18.7	3.05
UCFL-348	?	0.121	0.69	1.70	2.51	0.68	4.70	21.3	4.20
1712	O9	0.109	0.70	1.19	1.75	0.68	0.97	23.6	0.84
2203	O126	0.101	0.75	2.01	3.79	0.53	4.63	23.6	1.77
1763	O7	0.061	0.75	1.49	2.80	0.53	1.17	19.5	3.63
2105	O36	0.073	0.76	1.98	3.27	0.61	6.53	20.3	1.28
1367	O101	0.097	0.79	1.52	3.41	0.45	0.17	29.2	3.26
1852	O-autoagg.	0.090	0.81	1.92	3.14	0.61	2.93	21.7	7.73
2171	O149	0.031	0.83	1.60	2.36	0.68	0.77	18.1	1.48
2217	O174	0.048	0.84	2.08	3.54	0.59	5.17	24.5	2.42
1990	O54	0.072	0.85	1.94	3.41	0.57	3.23	20.6	1.63
2195	O21	0.049	0.86	1.94	3.52	0.55	1.47	23.1	0.14
2277	O175	0.048	0.88	1.71	2.96	0.58	3.87	23.1	2.85
2153	O117	0.033	0.89	1.85	3.50	0.53	2.70	24.3	2.36
1941	O15	0.038	0.91	1.66	3.52	0.47	1.13	26.3	0.11
2317	O-untypable	0.047	0.91	1.40	2.45	0.57	5.23	20.4	0.15
2280	O41	0.045	0.95	1.77	3.28	0.54	0.50	31.1	0.48
2262	O66	0.020	0.99	2.04	3.66	0.56	4.27	21.1	5.20
2606	O128	0.022	1.01	1.85	3.20	0.58	0.23	23.3	3.52
2134	O82	0.021	1.09	1.80	3.47	0.52	5.10	19.9	0.34
2637	O88	0.014	1.16	1.93	2.87	0.67	2.17	25.2	3.37
1625	O8	0.021	1.16	1.35	1.75	0.77	0.97	23.2	44.20
2059	O25	0.026	1.22	1.92	3.18	0.60	3.23	21.1	1.31
2052	O14	0.009	1.22	1.68	2.64	0.64	0.80	23.5	1.76
1935	O77	0.020	1.25	1.77	2.94	0.60	3.03	18.8	49.61
2041	O156	0.010	1.42	1.64	2.48	0.66	3.10	21.0	0.11
1514	O51	0.008	1.61	1.45	2.38	0.61	3.17	19.6	3.25
2049	O112	0.004	2.10	1.76	3.20	0.55	2.37	25.3	14.11

3.3.2 Sphericity, motility, zeta-potential, and aggregation

The relation between attachment efficiency and sphericity was very poor ($R^2 = 0.01$, $p = 0.48$; Fig. 3.2). The same was true for motility ($R^2 = 0.02$, $p = 0.30$), zeta-potential ($R^2 = 0.01$, $p = 0.53$). Data on aggregation were not normally distributed, therefore a non-parametric Spearman correlation was performed, which revealed that the correlation between cell-aggregration and attachment efficiency was not significant ($R^2 = 0.004$, $p = 0.67$). From this we concluded that these four parameters were not able to contribute in explaining inter-strain attachment efficiencies.

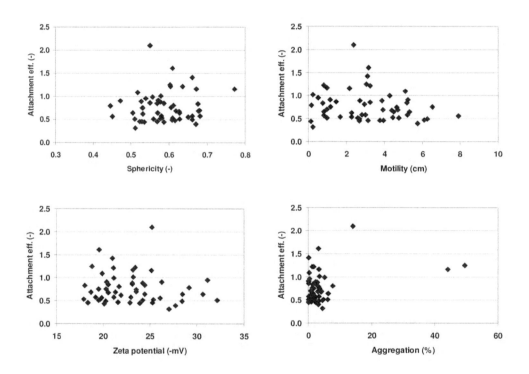

Fig. 3.2:   *Relation between attachment efficiency and the parameters cell sphericity, motility, zeta potential, and aggregation*

3.3.3 Serotypes and lipopolysaccharide structure

We grouped the various saccharides into 1. saccharides containing a carboxyl group or a phosphatyl group, 2. 'plain' saccharides, and 3. saccharides containing an amine group. The idea behind this grouping was that group 1 saccharides contain an acetate group or phosphate group, are acidic, and likely charged with a negative O-atom (Orskov et al., 1977; Stenutz et al., 2006), while group 2 and 3 saccharides are uncharged. The results are given in Fig. 3.3. We were able to derive the polysaccharide structure of 27 of the 54 strains we used. We observed that group 1 saccharides were more associated, but not statistically significant ($p = 0.13$ using Fisher's exact test), with low attachment efficiencies than with high attachment efficiencies: six of the nine strains with lowest attachment efficiencies had group 1 saccharides in their LPS structure, while for the 12 strains with highest attachment efficiency only one had a group 1 saccharide. Group 2 and group 3 saccharides were equally distributed and were not associated with attachment efficiencies ($p = 1$, Fisher's exact test).

Strain	Serotype	Attachment eff.	Glucuronic acid	Galacturonic acid	4-O-[(R)-1-carboxyethyl]-b-D-glucopyranose	O acetylated talose	O acetylated manose	3-O-[(R)-1-carboxyethyl]-L-rhamnose	Glycerol-1-phosphate	Rhamnose	Ribofuranose	Galactosefuranose	Mannose	Talose	Fucose	Galactose	Glucose	N-acetylgalactosamine	N-acetylglucosamine	N-acetylfucosamine	N-acetylmannosamine	N-acetylquinovosamine
			Group 1							*Group 2*								*Group 3*				
1923	O139	0.44	X							X								X				
2043	O29	0.45							X							X	X	X	X			
2007	O45	0.46					X										X		X			
2160	O124	0.46		X									X			X		X				
1991	O1	0.47								X							X			X		
2621	O5	0.49									X					X		X				X
2216	O58	0.51				X	X						X					X				
1876	O7	0.52								X			X			X		X				X
2312	O83	0.54	X													X	X	X				
1927	O16	0.54								X	X					X		X				
2000	O20	0.57									X					X						
1366	O23	0.64														X	X	X	X			
1875	O2	0.65								X								X	X			
2602	O159	0.68										X										
1712	O9	0.70				X										X		X				
2203	O126	0.75										X		X	X	X		X				
1763	O7	0.75								X		X			X		X				X
1367	O101	0.79																X	X			
2171	O149	0.83								X								X				
2195	O21	0.86													X	X	X	X			
2153	O117	0.89								X					X	X	X				
2262	O66	0.99	X										X	X			X				
2606	O128	1.01													X	X	X				
2637	O88	1.16											X	X			X				
1625	O8	1.16											X				X				
2059	O25	1.22								X						X		X	X			
1935	O77	1.25											X				X				

Figure 3.3: *Serotype, structure of the lipopolysaccharide, grouped into saccharides containing carboxyls and phosphate groups (Group 1), saccharides (Group 2), and saccharides containing an amine group (Group 3). Strains were ordered with increasing attachment efficiency*

3.3.4 Presence of genes associated with E. coli surface structures

Genes were screened by a PCR based method. Results are given in Fig. 3.4. Of all genes tested, the presence/absence of the fimbrial/afimbrial adhesin encoding gene *Afa* was most associated, but not statistically significant ($p = 0.21$ using Fisher's exact test), with attachment efficiency, since the gene was absent in three of the ten strains with low attachment efficiency, and present in all tested strains (10) with high attachment efficiency. The *fimH* gene of the type I fimbriae was present in all strains used in the PCR analysis, and so were the genes *ompC*, *slp*, and *surA*. Their presence was therefore not associated ($p = 1$, Fisher's exact test) with high or low attachment efficiencies.

	Strain	α	aah	AIDA-I	AIDA-C	Ag43 α	Ag43β	fimH	sfaS	focG	sfa/focDE	fanC	faeG	fedA	Afa	bmaE	Saa	nfaE	paa	iha	bfpA	ompC	slp	surA
low attach. efficiency	UCFL-94	0.32				X	X	X													X	X	X	X
	UCFL-131	0.40				X	X	X														X	X	X
	UCFL-71	0.40						X							X							X	X	X
	1923	0.44				X	X	X							X			X	X		X	X	X	X
	2007	0.46				X		X							X							X	X	X
	2160	0.46						X														X	X	X
	1991	0.47				X	X	X							X				X			X	X	X
	2621	0.49				X	X	X							X			X	X			X	X	X
	1924	0.50				X	X	X							X				X			X	X	X
	2216	0.51						X							X				X			X	X	X
high attach. efficiency	2606	1.01						X							X						X	X	X	X
	2134	1.09						X							X	X						X	X	X
	2637	1.16						X							X							X	X	X
	1625	1.16				X	X	X	X	X					X						X	X	X	X
	2059	1.22				X	X	X							X						X	X	X	X
	2052	1.22						X							X							X	X	X
	1935	1.25				X		X							X							X	X	X
	2041	1.42				X	X	X	X		X				X				X			X	X	X
	1514	1.61				X		X							X				X			X	X	X
	2049	2.10				X		X							X				X			X	X	X

Figure 3.4:   Results of the PCR analyses. A "X" indicates the presence of the gene. The strains used were grouped in 'low attachment efficiency' (upper ten strains) and 'high attachment efficiency' ('lower ten strains). Group A: autotransporter proteins, group B: fimbriae and adhesins, group C: pili, group D: other

The genes *flu*, *iha*, *bfpA* were present in, resp., 40%, 40%, and 25% of strains analyzed, and for all cases, there was no relation between presence/absence of the gene and attachment efficiency ($p = 0.65$, $p = 0.65$ and $p = 1$, resp.). Of course, the presence of a gene does not automatically indicate that the protein encoded by that gene is also expressed in the outer surface membrane of the *E. coli* strains, and therefore the conclusion that these genes were not involved in initial attachment was not justified. However, apparently, the presence of the genes was apparently not essential for the attachment of *E. coli* to the quartz grains. We considered the *flu* gene to be present, if both primer sets, amplifying different parts of the gene, yielded correctly sized PCR product. So, of the 20 strains considered for PCR, eight strains (UCFL-94, UCFL-131, 1923, 1991, 2621, 1625, 2059, and 2041) were positive for the *flu* gene. Besides the lack of relation between presence/absence of the *flu* gene and attachment efficiency, there was also no relation ($p = 0.67$; Kruskal-Wallis test) between presence/absence of the gene and cell aggregation.

S fimbriae and F1C fimbriae (*sfa/focDE* gene cluster, *focG*, and *sfaS* genes), the porcine associated gene (*paa*), the non-fimbrial adhesin (*nfaE*), and *bmaE* were present in (less than) 10% of the strains tested, while genes *aah, AIDA-I, AIDA-C, fanC, faeG, fedA,* and *Saa* were completely absent. We concluded that those genes amplified only once or twice with PCR analyses could not have played a significant role in initial attachment to the quartz sediment.

3.4 Discussion

Our work indicated that in a column of 7 cm in height, under identical flow conditions, there was a two log variation in maximum breakthrough concentrations of the 54 *E. coli* strains we used. Attachment efficiencies varied between 0.3 and 1, which was rather high, and in a number of cases, attachment efficiencies were above 1. When values were between 1 and 1.25, we did not consider those to be problematic. Values above unity might occur due to the fact that the TE correlation equation (Tufenkji and Elimelech, 2004a), which we used to determine the single collector contact efficiency, applies to spherical collectors. Our sand grains were somewhat rounded, not spherical. In such case, the single collector contact efficiency might be under estimated, giving rise to attachment efficiencies in excess of unity. Others (Shellenberger and Logan, 2002; Morrow et al., 2005; Paramonova et al., 2006; Yang et al., 2006; Lutterodt et al., 2009a) have also reported on attachment efficiencies above 1. For the remaining three strains (2041, 1514, and 2049), it was difficult to find an explanation for their very high attachment efficiencies of 1.42, 1.61 and 2.10, other than to assume that the bacteria size that was used in calculating the single collector contact efficiency had been wrong, most likely due to the presence of surface structures sticking out of the surface of these strains, thereby actually acting as a bigger colloidal particle than used in the TE correlation equation. As we indicated in the Methods Section, we wanted to create an environment inside the column which would enhance the attachment of cells. We succeeded in this (attachment efficiencies ranged between 0.3 and 1), because of the presence of divalent Ca and Mg in the solution: Foppen and Schijven (2006) reported similar attachment efficiencies for *E. coli* in suspensions of low ionic strength containing Ca and Mg. The two log variation in maximum breakthrough concentrations reinforced the idea that, although *E. coli* is an indicator organism, there is not one single attachment efficiency for all *E. coli* strains. There is variation, even at a very small scale. These variations are often neglected, especially when studying few *E. coli* strains (e.g. Foppen et al., 2007a,b; Tufenkji, 2006; Bradford et al., 2006; Walker et al., 2004; Powelson and Mills, 2001; Levy et al., 2007; Jiang et al., 2007).

If we try to establish the main drivers for these attachment efficiency variations, then sphericity, motility, zeta-potential, and aggregation were not helpful in explaining the inter-strain attachment variations. Lutterodt et al. (2009a), Bolster et al. (2009), and Walker et al. (2004) arrived at similar conclusions. However, these authors used a limited number of *E. coli* strains, while our sample size was much larger.

Other parameters apparently exerted greater control over the attachment efficiency. LPS occupies 75% of the surface of the bacterium, and *E. coli* is estimated to have 10^6 molecules per cell (Caroff and Karibian, 2003), so possibly one of the parameters exerting great control over the attachment efficiency could have been the presence of acidic carboxyl and phosphatyl groups in the LPS in the outer membrane of *E. coli*. These acidic groups are likely charged with a negative O-atom (Orskov et al., 1977; Stenutz et al., 2006). The quartz sand in our experiments was also negatively charged, giving rise to rather unfavorable conditions for attachment, on top of the fact that all strains already had a negatively zeta-potential. The LPS – attachment efficiency relation seemed somewhat confounded by other mechanisms, but at present we are unable to pinpoint these mechanisms. Except for Walker et al. (2004), who focused more on the role of the charge in the core LPS, a role of the O-antigen lipopolysaccharides in initial attachment of planktonic *E. coli* cells in saturated porous media has not been reported.

Of all genes tested, only the presence/absence of *Afa* was associated ($p = 0.21$) with attachment efficiency. The Afa/Dr adhesin is an essential adhesin in Diffusely Adhering *E. coli* (DAEC). DAEC have been identified from their diffuse adherence pattern on cultured epithelial cells, and they appear to form a heterogeneous group. The first class of DAEC strains includes *E. coli* strains that harbor Afa/Dr adhesins (Afa/Dr DAEC). These *E. coli* strains have been found to be associated with urinary tract infections (UTIs) (pyelonephritis, cystitis, and asymptomatic bacteria) and with various enteric infections (Servin, 2005). The role of *Afa* in the initial attachment to abiotic surfaces, like the quartz grains we used, has not been studied yet, and because of its association with the attachment efficiency, we think this role deserves more attention in the future.

The genes *fimH, ompC, slp* and *surA* were present in all strains analyzed. Of those, only Type I fimbriae (*fim*) are known to be involved in initial attachment. *E. coli* mutants lacking *fim* are dramatically defective in initial attachment to abiotic surfaces such as polyvinylchloride (PVC) under stagnant culture conditions in rich culture medium (Pratt an Kolter, 1998, in: Van Houdt and Michiels, 2005). In another study, a *fimA* mutant showed defective initial attachment in a biofilm assay conducted in minimal culture medium on submerged Pyrex slides used as substratum for biofilm growth in a continuous flow culture bioreactor (Beloin et al., 2004). Furthermore, increased expression of the *fim* cluster was observed in biofilms grown in minimal culture medium in continuous flow-chamber culture (Schembri et al., 2003), and Type 1 fimbriae appeared to be necessary for early biofilm formation on glass wool in rich culture medium (Ren et al., 2004). These observations illustrate that Type 1 fimbriae are critical for initial stable cell-to-surface attachment for several *E. coli* strains and in a range of different media and biofilm growing systems (Van Houdt and Michiels, 2005). Although we have no direct evidence of the presence of the fimbriae, due to its common presence, in our case, we assumed they were formed, and were likely involved in initial attachment. Possibly there was a relation between attachment efficiency and the number of type I fimbriae present at the surface of the *E. coli* strains. However, with the techniques we used we were unable to determine such relation.

The *ompC*, *slp*, and *surA* genes were present in 100% of the strains used, and the importance of these genes has been demonstrated during the initial stages of biofilm formation (Sauer, 2003; Otto et al., 2001; Prigent-Combaret et al., 1999; Tabe Eko Niba et al., 2007). However, as is true for most of the genes studied here, their role in initial attachment of planktonic *E . coli* cells is not clear.

The genes *iha*, *flu* and *bfpA* were present in many cases (25-80% of the strains analyzed), but their presence/absence was not associated with the attachment efficiency. The protein expressed by the *iha* gene has been shown to adhere to epithelial cells (Tarr et al., 2000), and to act as a virulence factor in *E. coli* O157:H7, but the role of *iha* in adhering to quartz grains is unknown. The same is true for the bundle forming pili gene *bfpA*: the protein expressed by the gene is known to adhere to biotic cells, but its role in initial attachment to abiotic cells is unknown. The role of Ag43 in relation to abiotic surfaces has been studied in more detail (Schembri et al., 2003; Schembri and Klemm, 2001, Yang et al., 2006), but mainly in relation to its increased importance during the initial stages of biofilm formation, after initial attachment has already taken place.

Finally, from this research a number of important issues emerge. Firstly, we concluded that attachment efficiency and LPS composition were associated. But what is then the quantitative relation between LPS composition and initial attachment of *E. coli* at the quartz grain surface? Do strains with similar serotypes have similar breakthrough behavior under all circumstances? Secondly, we identified the presence of a number of genes, and the presence of the *Afa* gene was most associated with the attachment efficiency, but with the techniques used, we were unable to quantify gene expression. In a next step of this research, the (quantitative) role of proteins expressed by the genes in initial attachment of *E. coli* cells should be studied. In addition, knock-out genes or isogenic mutants to confirm the role of the various surface structures may have to be used. Thirdly, have we included all relevant surface structures? Still around 38% of the more than 4200 different proteins encoded by *E. coli* are of unknown function (Madigan et al., 2009), and we might well have missed one or more genes involved in the initial attachment of planktonic *E. coli* cells. Micro-arrays to allow for the inclusion of other genes, that express surface structures we have not included in this research, are important and should be utilized. Finally, except for the UCFL strains, which were harvested from the soil of a pasture (Yang et al., 2006), all other strains were exclusively isolated from biotic environments. How relevant are these results from an environmental point of view? Or, conversely, which type of *E. coli* strains tend to travel farther, and which ones travel less far? Would it be possible to identify zones or groups of *E. coli* in aquifers, according to surface characteristics, type of geology, and type of hydrochemical environment? These are intriguing questions, which would assist in characterizing and assessing the fate of *E. coli* in the subsurface, both as an indicator organism and as a pathogen.

3.5 Conclusions

- In a column of 7 cm in height, under identical flow conditions, there was a two log variation in maximum breakthrough concentrations of the 54 *E. coli* strains we used.
- Of all factors we investigated (sphericity, motility, zeta-potential, cell-aggregation, lipopolysaccharide composition, and the presence/absence of 22 genes coding for different outer surface proteins) no single factor was able to explain attachment efficiency variations statistically significantly. However, low attachment efficiencies were associated ($p = 0.13$ using Fisher's exact test) with LPS containing saccharides with phosphate and/or carboxyl groups. These saccharide groups are acidic and likely charged with a negative O-atom (Orskov et al., 1977; Stenutz et al., 2006), which reduced attachment to the negatively charged quartz surface.
- Of the 22 genes tested, *Afa* was most associated ($p = 0.21$) with attachment efficiency. The role of *Afa* in the initial attachment to abiotic surfaces is however unknown, and because of its association with the attachment efficiency, we think this role deserves more attention in the future.

PART II

INTRA-STRAIN ATTACHMENT VARIATIONS AND THE
MINIMUM STICKING EFFICIENCY

Chapter 4 Determining the minimum sticking efficiency of six environmental *Escherichia coli* isolates

This chapter is based on:
G. Lutterodt, M. Basnet, J.W.A. Foppen and S. Uhlenbrook (2009): Determining minimum sticking efficiencies of six environmental Escherichia coli isolates. Journal of Contaminant Hydrology Vol. 110 p. 110-117.

Abstract

In health impact assessments, the sticking efficiency of a bacteria or virus population largely determines the transported distance of that biocolloid population, and hence, the potential health impact. However, at the same time, one of the most difficult parameters to estimate is the lower value of the sticking efficiency that should be used in calculating the health impact. In this paper, we introduce the concept of the minimum sticking efficiency (α_i) value of a bacteria population, including a method to determine the minimum sticking efficiency. Thereto, sticking efficiency distributions of 6 environmentally isolated *E. coli* strains were determined by carrying out laboratory column experiments over a transport distance of about 5 m. Experiments were conducted in de-mineralized (DI) water and in artificial groundwater (AGW). Sticking efficiencies were calculated for column segments (at varying distances from top of column) and fractions of total bacteria mass input in each segment were estimated by mass balance. The sticking efficiencies were highest close to the top of the column, near the point of bacteria mass input (0.103-0.352 in DI, and 1.034 – 9.470 for AGW) and reduced with distance with the lowest α_i values (10^{-5}-0.06 in DI and 0.006 – 0.283 in AGW) determined at the two most distant column segments (between 2.33 and 4.83 m from the top of the column). Power-law distributions best described the relationship between fraction of cells retained, F_i, and α_i. The minimum sticking efficiency was defined as the sticking efficiency belonging to a retained bacteria fraction of 0.001% of the original bacteria mass (total number of cells) flowing into the column ($F_i = 10^{-5}$), and coinciding with a 99.999% reduction of the original bacteria mass, and minimum sticking efficiencies were extrapolated from the fitted power law distributions. In the DI experiments, minimum sticking efficiency values ranged from as low as 10^{-9} (for *E. coli* strain UCFL-94) to 10^{-2} (for E. coli strain UCFL-348); in the AGW experiments, minimum sticking efficiency values ranged from 10^{-6} (for strain UCFL-94) to ≥ 1 (for strain UCFL-348). We concluded that in quantifying health impacts of biocolloids traveling in aquifers, the concept of the minimum sticking efficiencies, and the percentage of individual biocolloids of a total population having such low sticking efficiency, together with an inactivation rate coefficient, can serve as a useful tool to determine the maximum transported distance as a worst case scenario, and, hence, the potential health impact.

4.1 Introduction

Many waterborne disease outbreaks are caused by the consumption of groundwater contaminated by pathogenic microorganisms (Goss et al., 1998; Macler and Merkle, 2000; Bhattacharjee et al., 2002; Close et al., 2006). Pathogenic microorganisms find their way into the sub-surface through the spreading of sewerage sludge on fields, leakage from waste disposal sites and landfills (Taylor et al., 2004), or infiltration from cesspits, septic tank infiltration beds, and pit latrines. In situations where the distance between source of pollution and abstraction point is small, the risk of abstracting pathogens looms (Foppen and Schijven, 2006).

To predict the presence of pathogens in water, a separate group of microorganisms is usually used, generally known as fecal indicator organisms. Many microorganisms have been suggested as microbial indicators of fecal pollution (like enterococci, coliphages and sulphite reducing clostridial spores; Medema et al., 2003), but one of the most important indicators used worldwide is *Escherichia coli*. In a recent work, Schinner et al. (2010) reported different attachment efficiencies of five waterborne pathogens (Gram negative bacteria: *E. coli* O157:H7 ATCC 700927, *Yersinia enterocolitica* ATCC 23715, Gram positive bacteria: *E. Faecalis* ATCC 29212 and Cynobacteria. *M aeruginosa* UTCC 299 and *Anabaena flosaquae* UTCC 607) to quartz sand indicating that the heterogeneity in transport and attachment behavior observed among commensal strains (Bolster et al., 2009; Simoni et al., 1998; Albinger et al., 1994) may not differ from those of pathogenic strains. For long, prediction of microbial transport behavior in saturated porous has relied on the classical colloid filtration theory (CFT) by Yao et al. (1971). One of the characteristics of the theory is the use of the sticking efficiency, which is defined as the ratio of the rate of particles striking and sticking to a collector to the rate of particles striking a collector, and is mainly determined by electro-chemical forces between the colloid and the surface of the collector (Foppen and Schijven, 2006). According to the theory, the sticking efficiency is constant in time and place (Yao et al., 1971; Foppen et al., 2007; Tufenkji and Elimelech, 2004a). However, recent research results indicate that the sticking efficiency of a biocolloid population varies due to variable surface properties of individual members of the population, resulting in differences in affinity for collector surfaces (Albinger et al., 1994; Baygents et al., 1998.; Simoni et al., 1998; Li et al., 2004; Tufenkji and Elimelech, 2005a; Tong and Johnson, 2007; Foppen et al., 2007a,b). Based on these findings, an important question is: What type of distribution describes the variation in sticking efficiencies of a biocolloid population best? Some workers demonstrated that sticking efficiencies were distributed according to a power-law (Redman et al., 2001a,b; Tufenkji et al. 2003), while others found a log-normal distribution (Tufenkji et al., 2003; Tong and Johnson, 2007) or a dual distribution (Tufenkji and Elimelech, 2004b and 2005b; Foppen et al 2007a,b). However, all studies, aimed at revealing sticking efficiency distributions, have been conducted for very limited transport distances (centimeter to decimeter), and can therefore not be considered representative for longer transport distances, which are so important in microbial risk assessment of groundwater and therefore quantifying the potential health impacts of pathogenic microorganisms traveling in aquifers. In those health impact assessments, the <u>minimum</u> value of the sticking efficiency distribution, and the percentage of individual biocolloids of a total population having such low sticking efficiency, are crucial parameters, because the minimum value in combination with the amount of cells largely determine the maximum transported distance, and, hence, the potential health impact. Foppen and Schijven (2006) indicated that the range of sticking efficiencies of

Escherichia coli for geochemically heterogeneous sediment, based on a number of studies, ranged from 0.002 to 0.2.

Because of the importance of knowing the characteristics of a sticking efficiency distribution of a biocolloid population for long transport distances, the present work aimed at determining types of sticking efficiency distributions of 6 *Escherichia coli* strains and their minimum sticking efficiencies in relatively high columns (5 m) of saturated quartz sand. To enhance comparison with environmental conditions, we only used *E. coli* strains isolated from the environment. Furthermore, the strains were grown for environmentally realistic conditions, and the chemical quality of the bacteria suspensions we used was close to environmental conditions

4.2 Materials and methods

4.2.1 Bacteria growth and column experiments

Six *Escherichia coli* (*E. coli*) strains (UCFL-71, UCFL-94, UCFL-131, UCFL-167, UCFL-263 and UCFL-348) were obtained from the soil of a pasture used for cattle grazing (Yang et al., 2004), *E. coli* isolates were grown in an extract of filter-sterilized (mesh size: 0.45 μm) cow manure to mimic environmental conditions. To do this, fresh cow manure was collected from a farm (biological farm Ackersdijk, Delft), and stored at -20 °C in batches of 50 gram. Prior to every experiment, a batch of 50 gram cow manure was defrosted and mixed with de-mineralized (DI) water at a 1:20 ratio (EPA – 1312 Leach Method). Manure extraction was facilitated by acidifying the mixture to a pH of 5±0.05 with concentrated sulphuric acid and nitric acid at 60/40 weight percent mixture, and extraction was performed for 2 hours. The mixture was then centrifuged (IEC Centra GP 8- rotar 218/18cm) for 10 min at 4600 rpm, and then at 9000 rpm for 10 min (MSE high speed 18). The supernatant was sequentially filtered through a 0.45 μm and a 0.2 μm mesh size cellulose acetate membrane filter (47mm diameter). *E. coli* isolates were activated from a holy tube (pepton agar stock) in Luria Bertani Broth (DifcoTM LB Broth, Miller) for 6 hours at 37 °C while shaking at 120 rpm on an orbital shaker. The inoculum was then diluted 10^5 fold in the cow manure extract and incubated, while shaking on the orbital shaker at 120 rpm, for 72 hours at 21 °C until a stationary growth phase was reached at a concentration of ~10^8 cells/ml.

To study the distributions in sticking efficiency the *E. coli* strains, column experiments were conducted in demineralised (DI) water and in artificial groundwater (AGW). AGW was prepared by dissolving 526 mg/L $CaCl_2.2H_2O$ and 184 mg/L $MgSO_4.7H_2O$, and buffering with 8.5 mg/L KH_2PO_4, 21.75 mg/L K_2HPO_4 and 17.7 mg/L Na_2HPO_4. The final pH-value ranged from 6.6 to 6.8 and the Electrical Conductivity (EC)-value ranged from 1025 to 1054 μS/cm. The porous media comprised of 99.1% pure quartz sand (Kristall-quartz sand, Dorsilit, Germany) with sizes ranging from 180 to 500 μm, while the median of the grain size weight distribution was 356 μm. With this grain size, we excluded straining as a possible retention mechanism in our column: assuming a bacteria equivalent spherical diameter of 1.5 μm, the ratio of colloid and grain diameter was 0.004, which was well below the ratio (0.007) for which straining was observed by

Bradford et al. (2007) for carboxyl latex microspheres with a diameter of 1.1 mm suspended in solutions with ionic strengths up to 31mM (the ionic strength of the solutions we used was 4.7 mmol/L only). Total porosity was determined gravimetrically to be 0.40. Prior to the experiments, to remove impurities, the sand was rinsed sequentially with acetone, hexane and concentrated HCl, followed by repeated rinsing with de-mineralized water until the electrical conductivity was very low (<3 µS/cm).

The column consisted of a 5 m transparent perspex tube with an inner diameter of 10 cm, and with sampling ports placed at 10-50 cm intervals along the tube. A stainless steel grid for supporting the sand was placed at the bottom of the tube. The column was gently filled with the clean quartz sand under saturated conditions, while the sides of the column were continuously tapped during filling, to avoid layering or trapping of air. The column was connected both at the funnel shaped effluent end and influent end with two Masterflex pumps (Console Drive Barnant Company Barrington Illinois, USA) via teflon tubes, and the pumps were adjusted to a fluid approach velocity of 1.16×10^{-4} m/s, coinciding with flushing the column with 1 pore volume per working day. Prior to a column experiment, the column was flushed for two days with either DI or AGW to arrive at stable fluid conditions inside the column. Bacteria influent suspensions were prepared by washing and centrifuging (3000 rpm) three times in either DI or AGW, and then diluting 1000 times to arrive at bacteria cell concentrations of approximately 10^5 cells/mL. Experiments were conducted by applying a pulse of 0.15 PV (approximately 2.2L) of bacteria influent suspension to the column, followed by bacteria free DI or AGW. Samples were taken at 7 distances from the column inlet, and immediately plated (0.1 mL) in duplicate on Chromocult agar (Merck). Decay was assessed in all experiments by plating samples of the influent at an hourly frequency during the entire experiment. All plates were incubated at 37 °C for at least 18 hours.

After each experiment, to clean the sand in the column, and to prepare for the next experiment, a pulse of 5 L 1.9 M HCl followed by a pulse of 5 L 1.5 M NaOH was flushed through the column, followed by flushing with DI water until the electrical conductivity of the effluent was well below 3 µS/cm.

To determine auto-aggregation, 15 ml of freshly grown bacteria were centrifuged (14000 xg) and washed three times in AGW, and then allowed to stand for 180 minutes at a temperature of 4 °C. A sample of 1 mL 1cm below the surface of the suspension was obtained, immediately and 180 minutes after washing. The optical density of the samples was measured at 254 nm, and the auto-aggregation was determined as the ratio of the final over the initial optical density (in %).

4.2.2 Determining the sticking efficiency in each column slice

Crucial in assessing the characteristics of the sticking efficiency distribution of each *E. coli* strain, including determining the value of the minimum sticking efficiency, is the way in which the sticking efficiencies are calculated. Instead of considering the entire column length, we determined the sticking efficiency for each slice of column (Martin et al., 1996), in between two sampling ports:

$$\alpha_i = -\frac{2}{3} \frac{d_c}{(1-\theta)\eta_0 L_i} \ln\left(\frac{M_i}{M_{i-1}}\right) \qquad (4.1)$$

where α_i is the dimensionless sticking efficiency of column slice i, d_c is the median of the grain size weight distribution (m), η_0 is the single collector contact efficiency (-), θ is the total porosity of the sand (-), L_i is the height of the column slice i, i.e. the distance (m) between two sampling ports, M_{i-1} is the total number of cells entering slice i, obtained from the breakthrough curve determined at the upper sampling port of slice i and M_i is the total number of cells, obtained from the breakthrough curve determined at the lower sampling port of slice i. The total number of cells M in the fluid phase at a sampling port was obtained from (Kretzschmar et al., 1997)

$$M = q\int_0^t c(t)dt \qquad (4.2)$$

where q is the volumetric flow rate (mL/min), C is the cell concentration in the suspension (# cells/mL) and t is time (min). The Tufenkji-Elimelech correlation equation (Tufenkji and Elimelech, 2004a) was used to compute η_0. For this, we assumed that the bacteria density was 1055 kg/m^3, and the Hamaker constant was 6.5×10^{-21} J (Walker et al., 2004).

4.2.3 Retained bacteria as fraction of total input

For the assessment of the sticking efficiency distribution, not only the sticking efficiencies in all column slices need to be determined, but also the number of retained bacteria in a slice, as a fraction of the total number of bacteria cells injected in the column. This fraction, F_i, in each segment was calculated as

$$F_i = \frac{M_i - M_{i-1}}{M_0} \qquad (4.3)$$

where M_0 is the total number of cells in the influent.

4.2.4 Data analysis

Various types of distributions (logarithmic, exponential, power law) were used to assess the relation between the fraction of cells retained, F_i, and their corresponding sticking efficiency per column slice, α_i, thereby assessing the nature of the sticking efficiency distributions of the *E. coli* strains we used. To evaluate the goodness of fit, we employed the coefficient of determination, R^2. In case $R^2 > 0.9$, the fit was considered excellent. For $0.8 \le R^2 \le 0.9$, the fit was considered good, and when $R^2 < 0.8$, the fit was considered weak. All regression curve fitting were performed using SPSS 14 (SPSS, 2005).

4.3 Results

4.3.1 Breakthrough curves

An example of one experiment with 7 normalized breakthrough curves, measured at the 7 sampling ports at various distances from the column inlet is shown in Fig. 4.1 for UCFL-131 in DI (demineralised water).

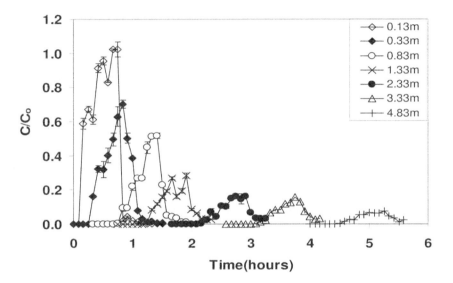

Fig.4.1:   *Breakthrough curve of UCFL-131 in DI. Error bars indicate variation between two duplicate plate counts and were determined by the difference between maximum and minimum plate count*

In general, peak concentrations reduced with transported distance, while both the rising and falling limbs of the breakthrough curves became less steep with transported distance. The latter was most likely due to increased dispersion, as the travel distance increased. Although normalized breakthrough concentrations for the other *E. coli* strains varied in maximum normalized concentrations, the trends in the breakthrough curves were similar (data not shown). Autoaggregation results (data not shown) indicated that, under the experimental conditions we employed, none of the strains had the ability to autoaggregate. We were therefore convinced that the number of culturable bacteria on Chromocult agar correctly reflected the number of culturable cells or the number of total cells measured during breakthrough.

4.3.2 Variation in sticking efficiencies

In DI, for all *E. coli* strains, α_i values in the top segments were highest (0.103-0.352; Table 4.1), and reduced with distance. Lowest α_i values (10^{-5}-0.06) were determined for the two most distant column segments (between 2.33 and 4.83 m from the top of the column). Negative values of α_i were computed for the third and fourth segments for UFCL-71 in DI.

Table 4.1: Results of α_i obtained in DI and AGW at distances from top of the column.

Distance from top of column to end of segment	Segment Thickness (m)	DI Water						AGW					
		UCFL-71	UCFL-94	UCFL-131	UCFL-167	UCFL-263	UCFL-348	UCFL-71	UCFL-94	UCFL-131	UCFL-167	UCFL-263	UCFL-348
0.13	0.13	0.352	0.103	0.140	n.d.	0.326	0.252	n.d	1.034	5.527	4.017	5.954	9.470
0.33	0.20	0.174	0.079	0.260	0.136	0.092	0.218	n.d	0.002	0.017	1.191	0.745	b.d.
0.83	0.50	-0.030	0.021	0.080	0.024	0.013	0.148	n.d	0.002	0.023	0.669	0.007	b.d.
1.33	0.50	-0.037	0.005	0.100	0.070	0.001	0.185	0.719*	0.002	0.047	0.139	0.126	b.d.
2.33	1.00	0.041	0.015	0.042	0.055	0.003	0.217	0.145	0.022	0.020	b.d	0.203	b.d.
3.33	1.00	0.003	0.027	0.020	0.090	0.032	0.168	0.019	0.020	0.283	b.d	b.d	b.d.
4.83	1.50	0.003	0.025	0.037	0.062	1.E-05	0.061	0.085	0.006	b.d	b.d	b.d	b.d.

* Value is for segment 0 -1.33 m, n.d., No data; b.d., Below detection limit

In these cases, the breakthrough at the end of the column segment was higher than at the beginning of the column segment. Although this phenomenon could possibly be due to release of attached cells, we were more inclined to attribute the negative values to inaccuracies in the method to determine the *E. coli* concentration (plating on Chromocult): in case of little attachment, the length of the column segment should be such that significant differences in *E. coli* influent and effluent mass can be measured in order to avoid negative values of α_i.

In AGW (artificial groundwater), α_i values in the top segments were highest (1.034 – 9.470), and reduced with distance. We did not consider sticking efficiency values in excess of 1 to be very strange. The presence of cell surface organelles like flagella and pili that extend beyond the cell surface may cause underestimation of the aspect ratio used in the computation of the single-collector contact efficiency (Shellenberger and Logan, 2002; Morrow et al. 2005; Paramonova et al. 2006). For many strains (UCFL-131, UCFL-167, UCFL-263, and UCFL-348), in AGW removal of *E. coli* was complete before the final sampling port was reached. In case of UCFL-348, removal was already complete within a traveled distance of 13-33 cm from the column inlet. For UCFL-71, *E. coli* breakthrough in sampling ports at 0.13, 0.33, and 0.83 m was not measured, and therefore, α_i was calculated for the segment 0-1.33 m (α_i = 0.719 in Table 4.1). Also in DI, negative α_i values were computed for UCFL-71, suggesting either a complex mechanism of release of attached cells or inaccuracies of the plating method, as was discussed above. For UCFL-94, the computed α_i values were invariably low, especially compared to the other strains. Overall, in AGW, for all *E. coli* strains we used, the computed α_i values for the more distant column segments ranged between $0.006 - 0.283$, and the order of magnitude of these low values (10^{-3}-10^{-2}) was similar to those observed in DI.

With eq. (4.3), for both the DI and AGW experiments, the fractions of retained bacteria over total bacteria mass in the influent suspension were computed, and the resulting distributions were plotted against the computed sticking efficiencies (Fig. 4.2a and b). On log-log scale, all distributions plotted more or less on a straight line for both the DI and AGW sets of experiments.

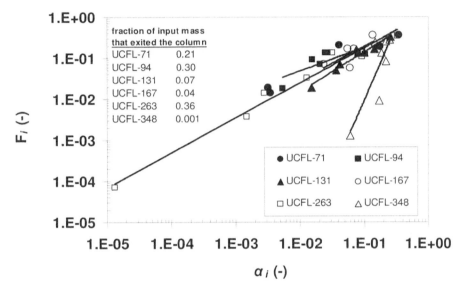

Fig.4.2a: *Retained bacteria, as fraction of input mass (F_i) and corresponding sticking efficiency for the DI experiments. Solid lines indicate fitted power law distributions.*

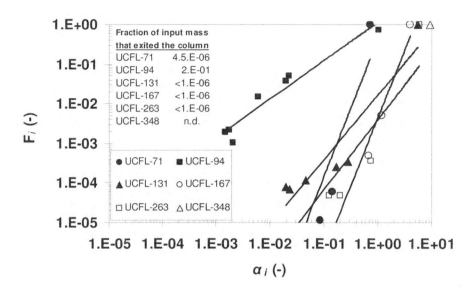

Fig. 4.2b: *retained bacteria, as fraction of input mass (F_i) and corresponding sticking efficiency for the AGW experiments. Solid lines indicate fitted power law distributions.*

Furthermore, in DI, the fractions were relatively close to each other (roughly between 0.001 and 0.3), while in AGW, there appeared to be a fraction close to 1 with a high sticking efficiency, and a very small fraction (roughly between 10^{-2} and 10^{-6}) with lower sticking efficiencies (0.1-0.001). Also given in Fig. 4.2a and 4.2b are the fractions of the total *E. coli* population leaving the column, without being retained. For the DI experiments, this fraction was 0.001 to 0.36; indicating that 0.1-36% of the bacteria mass in the influent suspension must have had an α_i value less than the lowest α_i values determined for the most distant column segments.

For the AGW experiments, the *E. coli* fraction leaving the column without being retained varied between less than 10^{-6} and 0.2 (UCFL-94), indicating that, generally, removal of the *E. coli* mass, with concentrations around $1-7\times10^5$ cells/mL was complete, while for one strain (UCFL-94) still 20% of the bacteria cells of the influent suspension must have had a sticking efficiency less than 0.001.

4.3.3 Sticking efficiency distributions

Exponential, power-law and logarithmic distributions were used to fit the relation between F_i and α_i (Table 4.2, Figures 4.2a and b). In DI water, R^2-values for all exponential distribution fits were below 0.8, ranging from 0.28 (UCFL-94) to 0.67 (UCFL-131). In addition, the R^2-values were statistically insignificant (p-values > 0.05) with the exception of UCFL-131 ($p = 0.02$) and

UCFL-348 (p = 0.03). The R^2-values for the curves fitted with a power-law distribution were significantly higher (0.55-0.98), and statistically significant, while R^2-values for the curves fitted with a logarithmic distribution were comparable to the exponential distributions, and, generally, statistically insignificant (p-values > 0.05).

In AGW, R^2-values for all exponential distribution fits were above 0.90, and, generally, statistically significant (p-values < 0.05). R^2-values for all power law distributions were also good ($R^2 \geq 0.83$) and statistically significant with the exception of UCFL-71(p = 0.09), while R^2-values for the curves fitted with logarithmical distribution were generally weak and statistically insignificant for most of the strains.

From this, we concluded that the power-law distribution described best the variations of α_i within the strains in both solutions, although the exponential distribution of α_i values was equally well capable of describing the distributions of α_i values in AGW.

Table 4. 2:   Coefficient of determination (R2) and probability (p) values of regression curve fitting of fraction of cells F_i, as a function of sticking efficiency, α_i

	DI Experiments						AGW Experiments					
	Exponential		Power		Logarithmic		Exponential		power		Logarithmic	
	R^2	p	R^2	p	R^2	p	R^2	p	R^2	p	R^2	p
UCFL-71	0.576	0.137	0.894	0.015	0.881	0.18	0.997	0.001	0.833	0.087	0.669	0.182
UCFL-94	0.276	0.226	0.630	0.033	0.636	0.032	0.590	0.044	0.937	0.000	0.819	0.005
UCFL-131	0.671	0.024	0.916	0.001	0.888	0.001	0.983	0.000	0.878	0.006	0.744	0.027
UCFL-167	0.643	0.055	0.553	0.090	0.562	0.086	0.900	0.051	0.977	0.012	0.581	0.237
UCFL-263	0.353	0.159	0.977	0.000	0.550	0.056	0.937	0.007	0.832	0.031	0.527	0.165
UCFL-348	0.637	0.031	0.696	0.020	0.354	0.159	n.d.	n.d.	n.d.	n.d.	n.d.	n.d.

n.d.: no data

4.3.4 Minimum sticking efficiencies

With the fitted power law distributions, we were able to extrapolate retained fractions and sticking efficiencies to values lower than the ones we determined in the 5 m column of quartz sand. For the sake of this paper, we assumed that when the retained bacteria fraction was reduced to 0.001% (5 log units) of the original bacteria mass (total number of cells) flowing into the column, i.e. $F = 10^{-5}$ in Figs. 4.2 a and b, then these retained cells possessed a so-called minimum sticking efficiency. The choice of a 5 log elimination was arbitrary in the sense that the total number of cells flowing into the column was much more than 10^5. However, because we injected a pulse of a constant concentration with constant velocity, the relation between bacteria mass and maximum bacteria concentration, along transport distance, was almost linear, assuming limited dispersion, as was the case in our experiments (Fig. 4.1). Therefore, we interpreted a bacteria mass reduction of 5 log-units as being equal to a bacteria concentration reduction of 5 log-units. We considered such a concentration reduction to be maximal for environmental conditions, since maximum *E. coli* concentrations in waste water are within the $10^4 - 10^6$ cells/mL range (Foppen and Schijven, 2006; Baxter and Clark, 1984; Canter and Knox, 1985). Minimum sticking efficiencies extrapolated in this way ranged from as low as 10^{-9} for UCFL-94 to 10^{-2} for UCFL-348 in the DI water experiments, and from 10^{-6} for UCFL-94 to ≥ 1 for UCFL-348 in the AGW experiments (Table 4.3).

Table 4. 3   *Fitted power-law equations between F_i and α_i and extrapolated minimum sticking efficiency values for a 5 log bacteria mass removal*

Bacteria Strain	Fitted power-law equation of measured data (DI)	Extrapolated minimum sticking efficiency (DI)	Fitted power-law equation of measured data (AGW)	Extrapolated minimum sticking efficiency (AGW)
UCFL -71	$F_i = 0.89\alpha_i^{0.7}$	8.50E-08	$F_i = 0.43\alpha_i^{3.49}$	4.70E-02
UCFL -94	$F_i = 0.62\alpha_i^{0.55}$	1.93E-09	$F_i = 1.15\alpha_i^{0.97}$	6.06E-06
UCFL-131	$F_i = 1.06\alpha_i^{0.92}$	3.45E-06	$F_i = 0.02\alpha_i^{1.63}$	9.44E-03
UCFL-167	$F_i = 1.34\alpha_i^{0.83}$	6.65E-07	$F_i = 0.004\alpha_i^{3.42}$	1.73E-01
UCFL-263	$F_i = 1.14\alpha_i^{0.82}$	6.81E-07	$F_i = 0.004\alpha_i^{1.77}$	3.39E-02
UCFL-348	$F_i = 38.5\alpha_i^{3.6}$	1.48E-02	-	≥ 1

4.4 Discussion

Our results showed that overall, in both DI and AGW, for all *E. coli* strains we used, the computed lower values of α_i were in the same order of magnitude (10^{-3}-10^{-2}). However, for the DI experiments, the fraction of *E. coli* mass that had passed the column without being retained over the *E. coli* mass in the influent suspension ranged between 0.001 to 0.36, indicating that 0.1-36% of the initial bacteria mass must have had an α_i value less than the lowest α_i values determined for the most distant column segments. For the AGW experiments, removal of the *E. coli* mass was complete, while for one strain (UCFL-94), still 20% of the bacteria cells of the influent suspension had a sticking efficiency less than 10^{-3}. We showed that the power-law distribution described best the variations of α_i-values within the strains in both solutions (DI and AGW), although in AGW, the exponential distribution of α_i values was equally well capable of describing the distributions of α_i values. Minimum sticking efficiencies, tentatively defined as the sticking efficiency belonging to a retained bacteria fraction of 0.001% of the original bacteria mass (total number of cells) flowing into the column ($F_i = 10^{-5}$), and coinciding with a 99.999% reduction of the original bacteria mass, were extrapolated from the fitted power law distributions. Minimum sticking efficiency values ranged from as low as 10^{-9} for UCFL-94 to 10^{-2} for UCFL-348 in the DI water experiments, and from 10^{-6} for UCFL-94 to ≥ 1 for UCFL-348 in the AGW experiments.

4.4.1 Sticking efficiency variations within and between E. coli strains

In both DI and AGW α_i varied from segment to segment, indicating differences in interactions between cells and the quartz grains. Large α_i values in the top segments of the column for all strains were attributed to removal of a stickier fraction of the population relative to other cells within the strains (Albinger et al, 1994; Baygents et al., 1998.; Simoni et al., 1998; Li et al., 2004; Foppen et al., 2007a,b). The good fit of the power law for all AGW experiments and three of the DI experiments (UCFL-71, 131 and 263) could be attributed to the comparatively high retention at the column inlet where stickier fractions were removed resulting in a wide variation of all α_i-values.

In AGW, the electrostatic repulsive barrier was reduced resulting in an increased attachment and fuelling significant attachment in the first segment and contributed to an even wider distribution in all α_i-values compared to the DI experiments. We think that the differences in α_i-values can be attributed to heterogeneity in cell population within the strains, due to variability in surface properties (Albinger et al.,1994; Baygents et al., 1998.; Simoni et al., 1998; Li et al., 2004; Tufenkji and Elimelech, 2005a,b; Tong and Johnson, 2007; Foppen et al., 2007a,b; Lutterodt et al., 2009a). From Yang et al. (2004) and Yang (2005), we know that the strains we used indeed have (surface) characteristics variations related to variations in zeta-potential, motility, hydrophobicity, and expression of an outer-membrane protein produced by the so-called antigen 43. This protein is thought to enhance the initial attachment of *E. coli* cells (Henderson et al., 1997), and was confirmed by our earlier work (Lutterodt et al., 2009a), in which we demonstrated that *E. coli* strains having the protein expressed at the outer surface membrane were stickier than strains without Ag43 protein. In that work we observed a reduction in the correlation of Ag43 expression and sticking efficiency along transport distance and the same observation was made for the relation between motility and sticking efficiency indicating a possibility of preferential removal of motile cells expressing the Ag43 adhesin. It can therefore be concluded that within a bacteria population, non-motile cells that neither express the Ag43 adhesin nor other surface characteristics that may facilitate cell adherence to the pure quartz grains are likely to possess the so called minimum sticking efficiency. Such variability in bio-colloid surface properties can result in an interaction potential distribution within the bio-colloid population (Li et al., 2004) leading to differences in cell-collector grain interactions, finally resulting in a distribution of α_i-values.

Geochemical heterogeneity on collector grain surfaces has been implicated as one of the factors leading to distributions in α_i and deviation of deposition patterns from the CFT (Johnson et al., 1996; Bolster et al., 2001; Loveland et al., 2003; Foppen and Schijven, 2005), but the 99.1% pure quartz grains we used in the experiments enabled us to exclude geochemical heterogeneity as a candidate for the observed differences in α_i-values. In addition, the possibility of straining as a contributing factor was ruled out, since the ratios of bacteria equivalent diameter to the grain diameter in the experiments were well below 0.007, as observed for the occurrence of straining (Bradford et al., 2007). The pulse application of bacteria solution also allowed us to eliminate blocking as the possible source of comparatively higher breakthrough at segments where α_i was negative.   It should be noted that all of the above explanations treat bacteria as 'static' biocolloids, unable to adapt to their environment. This 'unable to adapt'-concept, however, may not be true: subsurface transport and sticking efficiencies of chemotactic *Pseudomonas putida* G7 have, for instance, been found to heavily depend on the substrate availability and location in column experiments (Velasco-Casal et al., 2008), thereby demonstrating the effect of aut(ecological) adaptations. To our knowledge, information on relatively fast aut(ecological) adaptations of *E. coli* strains during transport in columns is not available in the literature, and the same is true for the relation between sticking efficiency variations and aut(ecological) adaptations. This could be an interesting topic for future research.

Results of the curve fitting exercise indicated that power-law and exponential distributions are very important in describing the probability distributions of the cells affinity for the quartz grains surfaces for the two solutions. Our results are consistent with observations made by Tufenkji et al. (2003) and Redman et al. (2001a,b) who observed power-law deposition patterns from the analysis of experimental results from other researchers and their experiments respectively. As explained earlier in this section the variation in cell surface properties of the strains results in differences in sticking efficiency and thus gives rise to the observed power-law probability distributions. Results obtained indicated that 64-99 % and 80-100 % respectively in DI and AGW of the cells affinity for quartz grain surfaces could be explained by a power-law distribution.

4.4.2 Minimum sticking efficiencies

Have we found a set of realistic values for the minimum sticking efficiencies, which are so important in quantifying health impacts of biocolloids traveling in aquifers? For ionic strengths comparable to groundwater conditions, including monovalent and divalent ions, and for biocolloid concentrations below 10^5 cells/mL, which we consider to be the maximum concentrations of pathogenic or fecal indicator organisms traveling in plumes of wastewater in aquifers or, more in general, saturated porous media, the minimum sticking efficiency for most of the strains we used was in the order of 10^{-2} or more (Table 4.3), while removal was complete within 5 m. This order of magnitude corresponded well with the values found by Foppen and Schijven (2006), who indicated that the range of sticking efficiencies of *Escherichia coli* for geochemically heterogeneous sediment, based on a number of studies, ranged from 0.002 to 0.2. However, one strain, UCFL-94, deviated from this general trend (minimum sticking efficiency = 6.06×10^{-6}). We believe that this deviation was due to differences in surface characteristics of UCFL-94 compared to the other strains, although we do not know what the differences were exactly. This deviation most likely showed the importance of surface characteristics in the initial attachment of *E. coli* cells, as was discussed above.

For the DI set of experiments, which we considered to be a worst case, with maximum transport of *E. coli* cells, the minimum sticking efficiencies were much lower (as low as 10^{-9}) than for the AGW set. In literature, we could not find another example that presented such low sticking efficiencies.

4.5   Conclusions

From the experimental results and observations the following conclusions can be drawn:

- In both DI and AGW, for all *E. coli* strains we used, the computed lower values of α_i from the column experiments were in the same order of magnitude (10^{-3}-10^{-2}). However, for the DI experiments, 1-36% of the initial bacteria mass must have had an α_i value less than the lowest α_i values determined for the most distant column segments.

- Our results showed that the power-law distribution described best the variations of α_i-values within the strains in both solutions (DI and AGW), although in AGW, the exponential distribution was equally well capable of describing the distribution of α_i values.

- Calculated minimum sticking efficiency values ranged from as low as 10-9 for UCFL-94 to 10^{-2} for UCFL-348 in the DI water experiments, and from 10^{-6} for UCFL-94 to ≥ 1 for UCFL-348 in the AGW experiments.

Chapter 5 Transport of *Escherichia coli* in 25 m columns

This chapter is based on:
G. Lutterodt, J.W.A. Foppen A. Maksoud and S. Uhlenbrook (2011) Transport of Escherichia coli in 25 m columns. Journal of Contaminant Hydrology 119, p 80-88

Abstract

To help improve the prediction of bacteria travel distances in aquifers laboratory experiments were conducted to measure the distant dependent sticking efficiencies of two low attaching *Escherichia coli* strains (UCFL-94 and UCFL-131). The experimental set up consisted of a 25 m long helical column with a diameter of 3.2 cm packed with 99.1 % pure-quartz sand saturated with a solution of magnesium sulfate and calcium chloride. Bacteria mass breakthrough at sampling distances ranging from 6 to 25.65 m were observed to quantify bacteria attachment over total transport distances (α_L) and sticking efficiencies at large intra-column segments (α_i) (> 5 m). Fractions of cells retained (F_i) in a column segment as a function of α_i were fitted with a power-law distribution from which the minimum sticking efficiency defined as the sticking efficiency of 0.001% bacteria fraction of the total input mass retained that results in a 5 log removal were extrapolated. Low values of α_L in the order 10^{-4} and 10^{-3} were obtained for UCFL-94 and UCFL-131 respectively, while α_i-values ranged between 10^{-6} to 10^{-3} for UCFL-94 and 10^{-5} to 10^{-4} for UCFL-131. In addition, both α_L and α_i reduced with increasing transport distance, and high coefficients of determination (0.99) were obtained for power-law distributions of α_i for the two strains. Minimum sticking efficiencies extrapolated were 10^{-7} and 10^{-8} for UCFL-94 and UCFL-131, respectively. Fractions of cells exiting the column were 0.19 and 0.87 for UCFL-94 and UCL-131, respectively. We concluded that environmentally realistic sticking efficiency values in the order of 10^{-4} and 10^{-3} and much lower sticking efficiencies in the order 10^{-5} are measurable in the laboratory. Also power-law distributions in sticking efficiencies commonly observed for limited intra-column distances (< 2 m) are applicable at large transport distances(> 6 m) in columns packed with quartz grains. High fractions of bacteria populations may possess the so-called minimum sticking efficiency, thus expressing their ability to be transported over distances longer than what might be predicted using measured sticking efficiencies from experiments with both short (<1 m) and long columns (> 25 m). Also variable values of sticking efficiencies within and among the strains show heterogeneities possibly due to variations in cell surface characteristics of the strains. The low sticking efficiency values measured expresses the importance of the long columns used in the experiments and the lower values of extrapolated minimum sticking efficiencies makes the method a valuable tool in delineating protection areas in real-world scenarios.

5.1 Introduction

In protecting groundwater sources (springs, wells, boreholes, etc.), used for consumption purposes, from contamination by pathogenic microorganisms, distances are set between potential sources of contamination and wells or springs (e.g. Taylor et al., 2004). Mostly, the objective is to obtain target concentrations or a specified degree of removal of the pathogenic microorganisms or indicator organisms over a given distance (Tufenkji et al., 2003). Traditionally, the classical colloid filtration theory (Yao et al., 1971, Tufenkji and Elimelech, 2004a) is applied to predict particle transport distances. The theory predicts a first-order reduction in both fluid-phase and retained colloid concentration with transport distance. First order deposition of colloids with distance is based on the assumption that colloid affinities for surfaces of collectors are invariable, resulting in a constant colloid sticking efficiency. However, often, the required reductions in bacterial concentrations at estimated target distances are not achieved. The failure of the filtration theory to accurately predict bacteria transport distances has been ascribed to variations in bacteria sticking efficiencies (Albinger et al., 1994, Tong and Johnson, 2007, Foppen et al., 2007a, b, Lutterodt et al., 2009a, b, Dong et al., 2006, Li et al., 2004, Redman et al., 2001, 2001a Albinger et al., 1994). Recent research has indicated that the cell affinity for collector surfaces varies within (Lutterodt et al., 2009a, b,chapters, 2&3) and among (Lutterodt et al., 2009a,b, chapters 2 and 3,Bolster et al., 2009, Foppen et al., 2010) bacteria strains. Recently, Foppen et al. (2010) reported 2 log unit variation in the maximum relative peak breakthrough concentrations of 54 *Escherichia coli* (*E. coli*) strains over 7 cm transport distance under similar physical and aqueous chemical conditions, confirming the inter-strain differences observed by many researchers. The intra-strain and inter-strain heterogeneities have necessitated the need for cautious extrapolation of experimental results obtained from a single strain to predict transport distances (Bolster et al., 2009, Lutterodt et al., 2009a) and the inappropriateness of using a single *E. coli* test strain as a fecal indicator for removal of other enteric pathogens (Yang et al., 2008).

An important question is how low the sticking efficiency of fractions of cells within a population can be?. Note that lower values may enable longer transport distances. Most experiments that revealed sticking efficiency distributions of biocolloid populations have been conducted over relatively short distances of 0.05 to 1 m (e.g. Foppen et al., 2007a,b, Brown and Abramson, 2006, Walker et al., 2004, Tufenkji and Elimelech, 2005b, Tufenkji, et al. 2003, Bolster et al. 2000, Martin et al., 1996). As a result, high sticking efficiency values in the order 10^{-2} to 1 were estimated, which indicate complete removal within very short distances. A review by Foppen and Schijven (2006) showed that the sticking efficiencies of *E. coli* determined from field experiments were lower (0.002-0.2) than those determined under laboratory conditions (0.02-0.9), possibly due to the presence of preferential flow paths and dissolved organic matter known to enhance particle transport in aquifers. Thus the application of laboratory measured values to predict real-world transport distances likely results in the underestimation of actual bacteria transport distances in the subsurface with consequences of polluting drinking water sources (springs, boreholes, and wells). Until now, the longest reported transport distances of any form of laboratory controlled colloid transport experiment has been conducted over 8 m (Close et al., 2006) for microspheres and 4.83 m (Lutterodt et al., 2009a,b) for *E. coli* with gravels and quartz sand as porous media, respectively.

In an effort to improve the prediction of colloid transport distances in the environment, we recently introduced the so-called minimum sticking efficiency (Lutterodt et al., 2009a,chapter 2), defined as the sticking efficiency belonging to a bacteria fraction of 0.001% of initial bacteria mass flowing into a column, after removal of 99.999% (5 log reduction) of the original bacteria mass has taken place. The minimum sticking efficiency practically represents the sticking efficiency of a minor fraction of bacteria cells. However, within this minor fraction of bacteria cells, the sticking efficiencies are again not a constant, but they are distributed, and therefore, within this sub-fraction, the minimum sticking efficiency is the *highest* possible sticking efficiency. Within a bacteria population, the fraction of cells possessing the minimum sticking efficiency are those lacking properties (e.g. cell motility, fimbriae, presence of antigen-43, an outer membrane protein of *E. coli*-Lutterodt et al., 2009a, chapter 2), that influence cell attachment to collector grain surfaces and therefore have a relatively good ability to be transported over longer distances. This minimum sticking efficiency can be obtained through extrapolations from equations generated from (power-law) distributions of fractions of cells retained and their corresponding sticking efficiencies. These distributions have not been studied for large intra-column distances over long transport distances (> 5 m).

In this paper, a new approach in conducting bacteria transport experiments in the laboratory is introduced. This involves the use of a flexible helix column packed with quartz sand saturated in water. The main objectives are to measure sticking efficiencies that can be considered as environmentally realistic, and to determine the minimum sticking efficiencies of two *Escherichia coli* strains over relatively long transport distances of 25 m.

5.2 Materials and methods

5.2.1 Column set up and tracer experiment

The porous media consisted of 99.1% pure quartz sand (Kristall-quartz sand, Dorsilit, Germany) with grain sizes ranging from 180 to 500 μm, and a median grain size of 356 μm. Though the removal of chemical impurities from the surface of quartz grains may expose physical imperfections like cracks, edges and lattice defects which may produce variation in the surface charge (Stumm and Morgan, 1996) and can increase the attachment efficiency of bacteria, the quartz grains was washed with acetone, hexane and concentrated HCl followed by repeated rinsing with de-mineralised (DI) water until the Elecetrical conductivity of the effluent was very low (< 3μS/cm). This was to ensure interaction between bacteria cells and pure quartz grains with negligible influence of polar organics and polar in-organics which may also affect bacteria attachment rates.

To form a helix column, a 26 m flexible tube reinforced with stainless steel (Armovin HNA-Food production, Italy) and of diameter 3.2 cm was sealed at one end with a 50μm nylon mesh. The column was filled with the cleaned quartz sand under saturated conditions, while the sides of the flexible tube were continuously tapped with a plastic hammer to avoid layering and or trapping of air. The saturated column was then coiled around a 2.33 m diameter circular flexible wooden frame to form a helix column with 3.5 revolutions (Fig. 5.1). The choice of a large diameter (2.33 m) and a negligible pitch is to reduce the curvature of the helix in order to reduce the effect of shear flow (Nield and Kuznetsov, 2004) associated with high curvature helical columns. In helical columns, the ratio of the radius to the sum of the square of the radius and the square of the pitch defines the curvature (Nield and Kuznetsov, 2004).

Sampling ports were placed on one side of the column, at distances of 6, 12.15, 19 and 25.65 m from the influent end of the column and each extended from the outer circumference to the axis of symmetry (1.6 cm). This was to ensure that samples were representative of breakthrough at the respective distances. The column was connected at the influent end with a Master flex pump (Console Drive Barnant Company Barrington, Illinois, USA) via a teflon tube. The pump discharge was adjusted to a fluid approach velocity of 1.14×10^{-4} m/s to coincide with a flushing regime of 1 pore volume (PV) in 24 hours.

An important advantage of using a helical column is the ability to conduct bacterial transport experiments over long distances within a relatively small laboratory space. Since attachment efficiencies are expected to reduce with increasing transport distances, the arrangement is applied to measure very low values which are very important in determining transport distances relevant for 'real case' situations in the environment. The disadvantage of a helical column for bacteria transport experiments is the variable flow velocity across the cross-section of the column with faster flow in the inner part, which results in a dispersion and stretched breakthrough of an injected solute (Benekos et al., 2006, Cirpka and Kitanidis, 2001). In order to assess the longitudinal dispersivity and the possible occurrence of shear flow within the system, tracer experiments were conducted. Thereto, the column was continuously flushed with de-mineralized (DI) water until the EC was equal to the EC of DI water. Then, the column was flushed with 0.8 L of 1.5 g/L NaCl solution with an EC of 2.87 mS/cm, while 2 mL samples were taken at the four sampling distances at 30 minutes intervals. Each sample was diluted 10 times in 18 mL DI water and the EC were measured using an EC meter.

Fig. 5.1: Set up of the helical column

5.2.2 Bacteria growth and column experiments

Two *Escherichia coli* (*E. coli*) strains (UCFL-94 and UCFL-131) were selected for their low sticking efficiency, based on the work of Foppen et al. (2010). The strains had originally been obtained from the soil of a pasture used for cattle grazing (Yang et al., 2004). To grow bacteria, the isolates were activated from a peptone agar stock in Luria Bertani Broth (DifcoTM LB Broth, Miller) for 6 hours at 37 °C, while shaking at 150 rpm on an orbital shaker. Then, 5 mL of the of the bacteria suspension was diluted in 250 mL of Nutrient Broth and incubated while shaking on the orbital shaker for 24 hours at 37 °C to reach a stationary growth phase at a concentration of ~10^9 cells/ml.

To study the distributions of sticking efficiencies and to measure very low sticking efficiency values of the *E. coli* strains, bacteria transport experiments were conducted using artificial groundwater (AGW). AGW was prepared by dissolving 526 mg/L $CaCl_2.2H_2O$ and 184 mg/L $MgSO_4.7H_2O$ in DI water, and buffered it with 8.5 mg/L KH_2PO_4, 21.75 mg/L K_2HPO_4 and 17.7 mg/L Na_2HPO_4. The resultant pH-value ranged between 6.6 and 6.8 and the Electrical Conductivity (EC)-value ranged from 1025 to 1054 µS/cm. Prior to a bacteria transport experiment, the column was flushed with AGW until the pH and EC at the effluent end of the column were the same as of AGW. Bacteria influent suspensions were prepared by washing and centrifuging (4500 rpm) three times in AGW and then diluting 10 times in 0.9 L AGW to obtain bacteria cell concentrations of 10^8 cells/mL. Experiments were conducted by applying a pulse of 1 L (approximately 0.124 PV) of bacteria influent suspension to the column, followed by bacteria free AGW. Samples were taken at 4 distances from the inlet of the column, and the optical density (OD) of each sample was determined by a spectrophotometer (Cecil 1021, Cecil Instruments Inc, Cambridge, England) at an absorbance of 254 nm. Values were then converted to cells/mL via a calibration curve between number of cells/mL and corresponding OD_{254} obtained from measurements of OD_{254} and corresponding number of cells obtain by plating on chromocult agar (Merck).

Due to the expected long duration of the experiments, bacteria inactivation experiments were conducted alongside column experiments. This was done by plating in triplicate 0.1 mL of 10^{-6} dilution of the influent on chromucult agar (Merck) at 2 hour intervals. All plates were incubated at 37 °C for at least 18 hours. After each experiment the sand in the column was cleaned by flushing the column with a pulse of 5 L 1.9 M HCl followed by a pulse of 5 L 1.5 M NaOH. The column was then flushed with AGW until the pH of the effluent was 6.8-7, and the EC-value was in the range of 1025 to 1054 μS/cm. To assess the possible presence of E. coli in the column prior to an experiment, samples were taken from all sampling ports and plated on chromocult agar followed by incubation at 37°C for 18 hours.

5.2.3 Determination of porosity

The porosity of the entire column was determined by cutting the saturated column into 1 m slices followed by extrusion of the sand into pre-weighed plastic containers. The sand was then dried in an oven for 24 hours. Porosities were determined from the mass of the slices, together with (known) bulk volume of a 1 m slice and the density of quartz sand (2.66 g/cm^3).

5.2.4 Sticking efficiency and segment sticking efficiency

Sticking efficiencies over total transport distances (α_L) (-) from the influent end of the column to a sampling port were computed as (Abudalo et al., 2005, Kretzschmar et al., 1997)

$$\alpha_L = -\frac{2}{3}\frac{d_c}{(1-\theta)L\eta_0}\ln\left(\frac{M_L}{M_0}\right) \tag{5.1}$$

where d_c is the median of the grain size weight distribution (m), η_0 is the single collector contact efficiency (-) , θ is the total porosity of the sand (-), and L is the travel distance (m). The Tufenkji Elimelech (TE) correlation equation (Tufenkji and Elimelech, 2004a) was used to compute η_0. Thereto, 1055 kg/m^3 was assumed for the bacteria density, while the Hamaker constant was estimated to be 6.5×10^{-21} J (Walker et al., 2004). M_0 is the total number of cells in the influent and M_L is the total number of cells in the effluent (-) at L obtained as (Kretzschzmar et al., 1997)

$$M_L = q\int_0^t C(t)dt$$
(5.2)

where q is the volumetric flow rate (mL/min), C is the cell suspension (# cells/mL) and t is the time (min).

To verify the distribution in cells affinity for collector surfaces, sticking efficiencies within column segments were computed for a fraction of cells retained in a column slice as (Lutterodt et al., 2009a, Martin et al., 1996).

$$\alpha_i = -\frac{2}{3}\frac{d_c}{(1-\theta)\eta_0 L_i}\ln\left(\frac{M_i}{M_{i-1}}\right)$$
(5.3)

where α_i is the dimensionless sticking efficiency of column slice i, L_i is the length of the column slice i, i.e. the distance (m) between two sampling ports, M_{i-1} is the total number of cells entering slice i, obtained from the breakthrough curve determined at the upper sampling port of slice i and M_i is the total number of cells, obtained from the breakthrough curve determined at the lower sampling port of slice i using equation (2). Also the number of retained bacteria in a slice, as a fraction of the total number of bacteria cells injected in the column was estimated. This fraction, F_i, in each segment was calculated as (Lutterodt et al., 2009b)

$$F_i = \frac{M_i - M_{i-1}}{M_0}$$
(5.4)

5.3 Data Analysis

5.3.1 Extrapolation of tracer breakthrough curves

Observed tracer breakthrough concentrations at 19 and 25.65 m were fitted with second order polynomial and the coefficient of determination (R^2) was applied to evaluate the goodness of fit. R^2-values greater than 0.9 were considered good. Equations generated from fitting were applied to extrapolate data for the incomplete section of the curves. Tracer recoveries were then computed from the extrapolated curves. The Pearson's mode skewness coefficient was applied to measure the skewness of tracer breakthrough curves.

5.3.2 Power-law distribution in sticking efficiencies and determination of minimum sticking efficiencies

We have previously shown that a power-law best described the distribution of bacteria cell affinity for quartz grain surfaces (Lutterodt et al., 2009a). Therefore, the power-law distribution was applied to assess the relation between F_i and their corresponding α_i by fitting. The coefficient of determination (R^2) was used to evaluate the goodness of fit. As indicated in the Introduction section of this paper, the minimum sticking efficiency (α_{min}) was defined as the sticking efficiency belonging to a bacteria fraction of 0.001% of initial bacteria mass flowing into a column, after removal of 99.999% (5 log reduction) of the original bacteria mass has taken place. However, within this minor fraction of bacteria cells, the sticking efficiencies are not constant, but distributed, and within this 0.001% sub-fraction, the minimum sticking efficiency is the **highest** possible sticking efficiency. The minimum sticking efficiency was extrapolated from a power-law equation, given as:

$$F_i = A\alpha_{min}^{\beta} \qquad\qquad\qquad (5.5)$$

where A and β are constants obtained from fitting the experimental data.

5.4 Results

5.4.1 Porosity determination

Porosities determined for 1 m column slices ranged from 0.35 to 0.41 with a mean of 0.37, and a standard deviation of 0.016. Porosities measured for all slices within a segment were averaged to obtain the porosity of each segment. As a result, each column segment had a porosity of 0.37 with the exception of the second segment (6.0 to 12.15 m; porosity = 0.38). From the low standard deviation of the mean, we concluded that the column was uniformly packed.

5.4.2 Tracer and bacteria breakthrough

Breakthrough curves (Figs. 5.2 and 5.3) obtained at the sampling ports show systematic reductions in peak relative concentrations with distance. Tests for asymmetry yielded low positive (0.84 at 6 m and 0.26 at 12.15 m) and low negative skewness (-0.83 at 19 m and -0.31 at 25.65 m) for the first two sampling ports and the two most distant sampling ports, respectively. Percentage mass recovery of tracer was comparatively higher at the first two sampling ports (90% for both distances) than at the two most distant sampling ports (75% for 19 m and 70 % for 25.65 m). R^2 values obtained by fitting curves with second order polynomial were good (0.96 for 19 m and 0.98 for 25.65 m), computed tracer recovery for extrapolated curves were 77% and 90 % at 19 m and 25.65 m, respectively.

To determine the dispersivities and flow velocities at the various sampling distances, tracer concentrations obtained at the sampling distances, together with measured porosities of the columns segments were simulated using HYDRUS 1-D (Šimůnek et al., 2008). For the fitting, we assumed that the column was straight standing and vertical, longitudinal dispersion was the only process causing spreading of solute concentrations, transport was one dimensional, and the salt tracer did not interact with the sediment. The HYDRUS model performance was high ($R^2 = 0.96$). Fitted dispersivities ranged from 10 to 90 cm. From the tracer results we concluded that the helical column was applicable to study particle transport using the colloid filtration theory normally applied for quantifying bacteria transport and attachment in one dimensional cases. Maximum relative breakthrough concentrations (C_{max}/C_0) for UCFL-94 reduced with increasing transport distance (Fig. 5.3). Reductions in C_{max}/C_0 at distances between 6 and 25.65 were small compared with the differences between the first and second sampling distances. The fraction of cells exiting the column was 0.19.

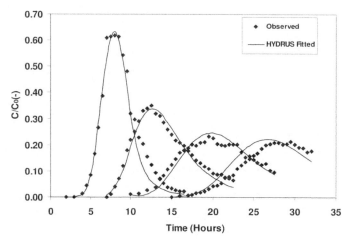

Figure 5.2:   *Measured breakthrough curve of a NaCl tracer, and the fit with HYDRUS 1-D*

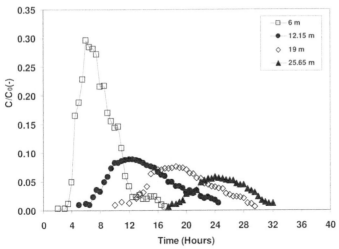

Fig. 5.3:   *Measured breakthrough curves of UCFL-94 at the various sampling ports (from 6 to 25.65 m*

Breakthrough concentrations for UCFL-131 were similar in shape and magnitude (curves not given). The fraction of cells exiting the column for UCFL-131 was 0.87. Inactivation experiments revealed little to no inactivation for both strains over the duration of the experiment (data not shown). Also, tests prior to experiments revealed no presence of *E. coli* after cleaning column.

5.4.3 Sticking efficiencies over total transport distances and segment sticking efficiencies

Measured sticking efficiencies α_L and segment sticking efficiencies α_i over total transport distances and column segments, respectively, indicated reductions in bacteria attachment with increasing transport distance. Over total transport distances α_L reduced with distance of transport (Fig. 5.4a) and varied from 4.55×10^{-3} at 6 m to 1.18×10^{-3} at 25.6 m for UCFL-94 and for UCFL-131 and ranged from 7.62×10^{-4} at 6 m to 2.26×10^{-4} at 25.65 m.

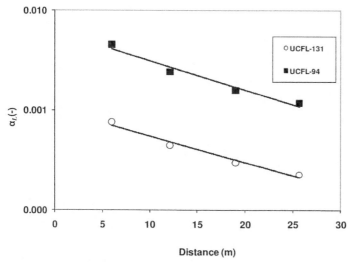

Fig 5.4a: Total sticking efficiencies for E. coli strains UCFL-94 and UCFL-131 as a function of transport distance

For the segments, α_i varied by less than 1 (0.65) log unit and by 2 log units for UCFL-94 and UCFL-131, respectively; values ranged from 8.22×10^{-6} to 4.55×10^{-3} for UCFL-94 and 1.55×10^{-5} to 7.62×10^{-4} from the effluent end to the influent end of the column. At equivalent transport distances values of α_L for the two strains revealed more than 0.5 log unit inter-strain variations (Table 5.1). From the results it can be concluded that very low values of sticking efficiencies are measurable in the laboratory using long columns.

Table 5.1:   Sticking efficiencies measured for total traveled distances (from influent end of the column) and segment sticking efficiency, measured for column segments

Distance from influent end of column	Sticking efficiency over total distance (α_L)		Segment thickness (m)	Segment sticking efficiency (α_i)	
	UCFL-94	UCFL-131		UCFL-94	UCFL-131
6.00	4.55E-03	7.62E-04	6.00	4.55E-03	7.62E-04
12.15	2.44E-03	4.47E-04	6.15	3.85E-04	1.42E-04
19.00	1.59E-03	3.00E-04	6.85	9.09E-05	3.99E-05
25.65	1.18E-03	2.26E-04	6.65	8.22E-06	1.55E-05

5.4.4 Power-law distribution in segment sticking efficiencies and extrapolated minimum sticking efficiencies

The relationship between F_i and corresponding α_i-values was determined by a power-law distribution (Fig. 5.4b): each fit had a R^2 value of 0.99 or higher. With the power-law distribution equations the minimum sticking efficiencies of the two bacteria strains were determined: 10^{-7} for UCFL-94 and 10^{-8} for UCFL-131(Table 5.2).

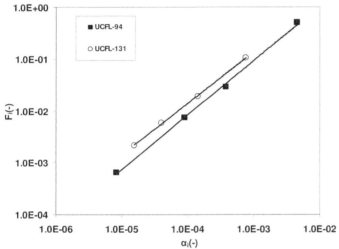

Fig 5.4:   Fraction of cells retained (F_i) as a function of segment sticking efficiency (α_i) for the two E. coli strains. Fractions exiting the column are respectively 0.19 and 0.87 for UCFL-94 and UCFL-131, respectively

Table 5.2: Fitted power-law equations between (F_i) and extrapolated minimum sticking ($\alpha_{i\,min}$)

Strain	Power-law equation	Coefficient of determination (R^2)	($\alpha_{i\,min}$)
UCFL-94	$F_i = 129.8\alpha^{1.05}$	0.99	1.68E-07
UCFL-131	$F_i = 125.69\alpha^{0.99}$	0.99	6.75E-08

5.5 Discussion

Low values of total sticking efficiencies in the order 10^{-4} and 10^{-3} were measured for UCFL-131 and UCFL-94, respectively, while values of segment sticking efficiencies were as low as 10^{-6} to 10^{-5}. Both α_L, the total sticking efficiencies and α_i, the sticking efficiencies determined for column segments, decreased with increasing transport distance. Minimum sticking efficiency values were extrapolated to 10^{-8} and 10^{-7} for UCFL-131 and UCFL-94, respectively. Power-law distributions could be used to describe the distribution of sticking efficiencies of bacteria strains at large travel-distances (> 6m), whereby substantial fractions (>10%) of bacteria mass were transported over distances in excess of 25 m.

5.5.1 Tracer and bacteria breakthrough

Though tracer breakthrough curves looked symmetrical, a test for skewness of the curves revealed low positive and low negative at the first two sampling ports and the two most distant sampling ports, respectively. The observed low positive skewness indicated possible tailing of the breakthrough and could be attributed to the possible settling of sand particles, which might have caused differential porosities across the vertical face of the column and promoted the possible occurrence of transverse dispersion, in spite of the uniform porosities measured at the column segments. The measured low negative skewness is mainly due to the incomplete nature of tracer breakthrough curves; this assertion is supported by the less than 100 % mass recoveries computed for all sampling distances and even for extrapolated breakthrough, which confirms the possibility of tailing. In spite of these limitations, there was negligible effect of shear flow in the set up as indicated by the high performance of the HYDRUS-1D and can be attributed to the large diameter of the helix formed, which reduced the curvature of the entire set up and minimized the effect of shear flow. It is known that higher curvature helix columns induce variable flow across the cross section thus increasing the axial velocity in one half compared to the other (Nield and Kuznetsov, 2004, Benekos et al., 2006). In addition to the possible contribution of differential porosities to high bacteria breakthrough, the long duration of the experiments might have caused starvation of the cells with possible changes in important surface properties causing low adhesiveness of the cells to the quartz grains (Haznedaroglu et al., 2008). Although there are limitations associated with the plating method of quantifying bacteria, such as the development of clumps and chain of cells into a single colony, and also the growth of only colonies for which cultural conditions are suitable for, results of inactivation tests indicated insignificant reduction in influent concentration over the duration of the experiments for both strains. We were therefore convinced that breakthrough concentrations measured indirectly via the spectrophotometer were indeed for culturable cells.

5.5.2 Low sticking efficiencies, inter-strain and intra-strain heterogeneities

Contrary to the classical colloid filtration theory, the common finding of variable *E. coli* attachment efficiency during transport in saturated porous media (Schinner et al., 2010, Bolster et al., 2009, Bolster et al., 2006) was also observed in our experiments. Sticking efficiencies reduced with increasing transport distances for the two bacterial strains we used, as was also observed by other workers (Lutterodt et al., 2009a, b Tufenkji and Elimelech, 2005b, 2004b, Simoni et al., 1998, Baygents et al., 1998). The low sticking efficiency values obtained in this work express the importance of the use of sufficiently long columns in bacterial transport studies, as the possibility of obtaining these values is not possible with short columns. Note that most values measured in laboratory setups reported in the literature range between 0.02 to 0.9 (Foppen and Schijven, 2006), and in some cases values greater than 1 were measured (Lutterodt et al., 2009a,b, Paramanova et al., 2006, Morrow et al, 2005, Shellenberger et al., 2002). These values are 2 to 5 orders of magnitude higher than results obtained in this work. The overall very low α_L values (in the order of 10^{-4} to 10^{-3}) we obtained are only comparable to the lower limits of reported field values (0.002-0.2; Foppen and Schijven, 2006). To our knowledge segment sticking efficiency (α_i) values as low as 8.22×10^{-6} obtained for UCFL-131 have not been reported in literature. Actual sticking efficiency values in the environment can even be much lower due to the ubiquitous presence of structural discontinuities and dissolved organic matters which are known to enhance particle transport in aquifers.

The differences in α_i and α_L measured for both strains confirm the intra-population and inter-population heterogeneities reported by other workers (Foppen et al., 2010; Bolster et al., 2009; Bolster et al., 2000, Baygents et al., 1998, Simoni et al., 1998, Tong and Johnson, 2007). The observed high values of both α_L and α_i at comparatively short distances of transport from the influent end of the column and decreasing with increasing transport distance is a phenomenon that has commonly been attributed to the removal of highly attaching cell fractions within bacterial populations (Lutterodt et al., 2009a,b, Foppen et al., 2007, Li et al., 2004, Albinger et al., 1994, Bolster et al., 1999, 2000) and can be explained by the variation in cell surface properties. A previous test for cell auto aggregation ability within the two strains under similar growth and hydrochemical conditions showed very low aggregation ability (<5 %) (Foppen et al., 2010). Therefore, the contribution of cell autoaggregation to the high sticking efficiency recorded at short transport distances was ruled out. Until now no cell property has been proved to convincingly influence *E. coli* attachment to quartz grains though cell motility and an auto transporter protein, Antigen-43 (Ag43) have been proved to influence cell attachment to quartz grains (Lutterodt et al., 2009a At equivalent transport distances the 5 times higher attachment of UCFL-94 than UCFL-131 may possibly be due to differences in cell properties that influence transport or attachment within the two strains. Previous work with these strains (Foppen et al., 2010, Lutterodt et al., 2009a, Yang, 2005) indicated differences in various cell properties (i.e. Ag43 expression, motility, outer surface potential, aggregation, cell size and sphericity) within and among the strains. For α_i, though the column segments were similar in length (6-6.85 m) considering the total length of the columns (>25 m), values varied indicating that α_i is not segment length dependent but dependent on total distance of transport. This observation is supported by the low correlation between α_i and L_i (data not shown).

5.5.3 Power-law distribution and minimum sticking efficiency

The observed power-law distributions in α_i are in agreement with observations made by other researchers (Lutterodt et al., 2009b, Brown and Abramson, 2006, Tufenkji et al., 2003, Redman et al., 2001, 2001a,). The large variations in α_i give rise to the power-law distribution (Lutterodt et al., 2009b) and may be due to variations in cell surface properties giving rise to differential affinity of cells for the collector grain surfaces as explained by others (Simoni et al., 1998, Baygents et al., 1998). The ratio of average cell size to the mean quartz grain size was less than 0.007 which seems too low for straining (Bradford et al., 2007), and the possibility of straining as a retention mechanism in the set up was therefore considered negligible.

The number of cells exiting the column shows that a substantial number of cells possess both α_i and α_L values which are lower than the minimum values measured in our experiments, and expresses the importance of the extrapolated minimum sticking efficiencies and also the length of the column used in experiments.

Minimum sticking efficiency values extrapolated from the power law equations were very low (10^{-7} and 10^{-8}) and the number of cells (10^6) possessing this so-called minimum sticking efficiencies indicates that a substantial fraction of bacterial sub-populations may be transported over very long transport distances. We have previously (Lutterodt et al., 2009b) observed that cells possessing the minimum sticking efficiency are non-motile fractions that lack the expression of Ag43 and other cell properties that may promote bacterial attachment to quartz grains. The low minimum sticking efficiency values extrapolated compared to values obtained in a previous work with the two strains (Lutterodt et al., 2009b) may be due to differences in growth conditions employed in the two experiments. The growth condition employed in this work (Nutrient broth at 37°C) might have caused low expression of cell surface properties that influences cell attachment to abiotic surfaces, causing low cell retention by the quartz grains. The two bacteria growth conditions, in the present work and in Lutterodt et al. (2009b) (cow manure extract at 21°C) are comparable to the intestinal and external environmental conditions, respectively, mimicked by Yang et al. (2006, 2008) who reported significantly higher expression of cell properties (e.g. Ag43 expression , hydrophobicity and biofilm formation) and higher retention on bio-barriers (Yang et al., 2008) for a number of *E. coli* strains grown under external environmental conditions than when grown under intestinal conditions.

5.5.4 Environmental implications

The low values of sticking efficiencies measured, fraction of cells exiting the column and the extrapolated minimum sticking efficiencies coupled with the number of cells possessing the minimum sticking efficiencies, have important implications for protecting water abstraction wells in aquifers. The minimum sticking efficiency gives the α_i-values of lower end fractions of cells in the power-law distribution with non-attaching characteristics. Taking into consideration factors that affect transport and attenuation of microorganisms in the environment (for e.g. preferential flow paths, presence of dissolved organic matter and metal oxides), such cells may be transported over distances much longer than predicted, if results from short column experiments are extrapolated to the field scale. The low values of extrapolated minimum sticking efficiencies compared with measured values of sticking efficiencies over the total transport distances and at column segments make the minimum sticking efficiency a valuable tool in delineating well-head protection areas in real-world scenarios.

5.6 Conclusions

The transport of two *Escherichia coli* strains was studied over 25 m using helical columns. Column segment sticking efficiency and sticking efficiency over total transport distances were determined. Fractions of total bacteria mass flowing into the column that were retained in column segments as a function of their corresponding segment sticking efficiency revealed power-law distributions in cell affinity for quartz grains. Also, a substantial sub-population of total input mass exited the column. The results in this study indicated that:

- Low values of sticking efficiencies in the order 10^{-3} to 10^{-6} are measurable in the laboratory, and the results demonstrate the importance of the use of long columns.
- Power-law distributions in sticking efficiencies commonly observed for limited intra-column distances (< 2 m) are applicable for the description of distributions at large segments (> 6 m) of columns packed with quartz grains.
- The power-law distribution in segment sticking efficiencies can provide useful information about transport of bacteria in the environment. In addition, it may provide a useful framework within which the transport and attachment of bacteria may be evaluated.
- Substantial numbers of bacteria cells may possess the minimum sticking efficiency indicating the possibility of being transported over distances much longer than may be predicted using measured sticking efficiencies from experiments with both short and long columns.

PART III

TRANSPORT OF *ESCHERICHIA COLI* STRAINS
ISOLATED FROM TERMINATION POINTS OF
GROUNDWATER FLOW LINES

Chapter 6 Transport of *Escherichia coli* strains isolated from spring water

This Chapter is based on:
G. Lutterodt, J.W. A. Foppen and S. Uhlenbrook (2011): *Transport of Escherichia coli strains isolated from spring water. Submitted to Journal of Contaminant Hydrology.*

Abstract

Although the transport of *Escherichia coli* (*E. coli*) strains in saturated columns of sediment was studied many times, to our knowledge, there is no study reported in literature that focuses on the transport of *E. coli* strains isolated *after* they have undergone transport in an aquifer. In this study, we harvested 6 *E. coli* strains from springs in Kampala, the capital of Uganda, and carried out laboratory column experiments with 1.5 m high quartz sand columns. To characterize transport, we used the minimum sticking efficiency (α_{min}). The results indicated that α_{min} for most of the strains was in a narrow range of minimum sticking efficiencies in the order of 10^{-5} to 10^{-4}. Based on these results, we propose that for worse case scenarios in predicting pathogen transport in aquifers, sticking efficiency values of 10^{-4} to 10^{-5} should not be considered unrealistically low. Instead, we demonstrated that these values are likely characteristic for most of the *E. coli* strains that would have undergone transport through an aquifer.

6.1 Introduction

Globally, groundwater systems provide 25–40% of the world's drinking water (Morris et al., 2003). The importance of the resource is often attributed to the assumption that it is free of pathogenic microorganisms (e.g. Bhattacharjee et al., 2002). However, many water borne disease outbreaks are caused by the consumption of groundwater contaminated by pathogenic microorganisms (Close at al., 2006; Bhattacharjee et al., 2002; Macler and Merkle, 2000; Powell et al., 2003).

Traditionally, strategies employed to protect groundwater sources from contamination rely upon effective natural attenuation of sewage-derived microorganisms by soils (and rocks) over set back distances (Taylor et al., 2004). The prediction of transport distances of microorganisms in aquifers has usually been determined with the classical colloid filtration theory (CFT; Yao et al., 1971; Tufenkji and Elimelech, 2004a, b). The theory is based on the assumption that colloid retention follows an invariable rate deposition on collector surfaces, while fluid phase colloid concentrations reduce log-linearly with increasing distance of transport. However, recent research results indicate that the sticking efficiency of a biocolloid population varies due to variable surface properties of individual members of the population, resulting in differences in affinity for collector surfaces (Albinger et al., 1994; Baygents et al., 1998; Simoni et al., 1998; Li et al., 2004; Tufenkji and Elimelech, 2005a; Tong and Johnson, 2007; Foppen et al., 2007, Lutterodt et al., 2009a). Like other workers (Redman et al., 2001a, b; Tufenkji et al. 2003), we demonstrated in our previous works (Lutterodt et al., 2009b, 2011) that a power-law best describes the distribution of relative bacteria mass fraction retained in the saturated porous medium and their corresponding so called segment sticking efficiencies (Lutterodt et al., 2009b, 2011) when transported through columns of saturated quartz sand. Others found a log-normal distribution (Tufenkji et al., 2003; Tong and Johnson, 2007) or a dual distribution (Tufenkji and Elimelech, 2004b, 2005a,b; Foppen et al., 2007). Based on these power-law distribution functions mentioned above, the segment sticking efficiency of 0.001% of the initial bacteria mass applied to a column was defined as the minimum sticking efficiency (Lutterodt et al., 2009b, 2011). This parameter quantifies the rate of interaction of a lower end non-attaching fraction of a bacteria population with clean, saturated, quartz sand. In any bacteria population, the fraction of cells that possesses the minimum sticking efficiency is the fraction that consists of cells with surface characteristics that promote transport or reduce the efficiency of attachment (Lutterodt et al., 2009b, 2011).

Escherichia coli (E. coli), a gram-negative, facultative non-spore forming, rod shaped bacterium is commonly used as indicator of fecal contamination of drinking water supplies, because *E. coli* is a consistent, predominantly facultative inhabitant of the human gastrointestinal tract. In addition, *E. coli* is easy to detect and quantify. Furthermore, the net negative surface charge and low inactivation rates of *E. coli* ensure that they may travel long distances in the subsurface and these characteristics make *E. coli* a useful indicator for fecal contamination of groundwater (Foppen and Schijven, 2006). Due to the importance of *E. coli*, considerable attention has been given to understanding their transport and fate in saturated porous media (e.g. Foppen et al, 2006, Schinner et al., 2010, Bolster et al., 2010). In most of these studies, bacteria strains isolated from different sources, for example from zoo animals (Foppen et al., 2010), a swine lagoon (Bolster et al., 2010), a dairy cow manure and sewerage (Haznedaroglu et al., 2008), or a soil of a pasture used for cattle grazing (Foppen et al., 2010, Lutterodt et al., 2009a, b, Yang et al., 2008) were used. To our knowledge, there is no study reported in the literature that focuses on the transport of *E. coli* strains isolated *after* they have undergone transport in an aquifer. These strains may start transport among a wide variety of microorganisms that have infiltrated from the surface into the groundwater system, and, after considerable transport has taken place, they may end up as one of the few microorganisms remaining in the system.

In this study, we hypothesized that if the segment sticking efficiency reduces with transport distance, as we have seen in many laboratory studies in literature, , then the minimum sticking efficiency of a number of *E. coli* strains harvested from natural springs (or: termination points of flow lines, when the transport distance is likely to be extended) is also expected to be low.

6.2 Materials and methods

6.2.1 The study area

Kampala is underlain by a variety of meta-sedimentary rocks and weathering has produced a pronounced topography. The town has a shallow aquifer in the weathered regolith and site investigations showed the presence of preferential flow paths, although it is not known how far these extend (Howard et al., 2003). Previous studies in Uganda indicated that the top of the regolith is composed of fine material, with increasingly coarser (sandy clay) material found at depth with productive aquifers commonly associated with these layers (Howard et al., 2003; Kulabako et al., 2007). Recharge tends to occur during two distinct wet seasons, although Kampala experiences rainfall throughout the year due to its proximity to Lake Victoria (Nyenje et al., 2010). Kampala has many springs, and usually these springs are protected with a concrete slab and a small protected area upstream of the spring (Fig. 6.1). In Kampala, six *E. coli* strains were isolated from springs similar to that indicated in Fig. 6. 1. They were located in Formal Residential areas (*E. coli* strains named FR02, FR05, and FR08), an urban FArm (FA03) and informal residential areas or SLums (SL03, SL20).

Sampling was undertaken in July 2010. Thereto, 250 ml spring water was collected in sterile polypropylene bottles, by means of a syringe 100 ml of each sample was passed through 0.45 μm cellulose acetate filter, the filter is then placed on chromucult agar (DifcoTM LB Broth, Miller) plate and transported to the microbiological laboratory of Makerere University and incubated at 37°C for 24 hours. The purple color of *E. coli* cells on chromucult agar plates allowed us to detect the *E. coli* among different bacteria species in growing on the agar plates. A sterile toothpick was used to pick a single colony of *E. coli* from agar plates and inoculated into 5 ml of Nutrient broth followed by incubation at 37°C for 24 hours. 1-2 ml of freshly grown *E. coli* cells were then inoculated into sterile vials (Microbank™-Dry, PRO-LAB DIAGNOSTICS, Toronto, Canada) containing porous beads saturated with cryopreservative (cryovials), which serve as carriers to support microorganisms. The vials are then stored at -70°C and then transported to the UNESCO-IHE laboratory, Delft, the Netherlands.

Fig 6.1:   *Example of a spring inside the town of Kampala, the capital of Uganda.*
               Spring water is used for drinking, cooking and bathing. Please note the
               concrete slab and the fencing behind the spring in order to try to protect the
               intake area

6.2.2 Bacteria growth and size measurements

To conduct experiments, a sterile forceps was used to pick a bead from the vial and into 25 ml of nutrient broth and then incubated for 24 hours at 37 °C while shaking at 150 rpm on an orbital shaker. Then, 5 ml each of the bacteria solution was further inoculated into four Erlenmeyer flasks containing 250 mL of nutrient broth and again grown for 24 hours on an orbital shaker at 150 rpm for 24 hours to obtain a cell concentration of $\sim 10^9$ cells/ml. Bacteria were washed and centrifuged (4600 rpm) three times in Artificial Ground Water (AGW), which was prepared by dissolving 526 mg/L $CaCl_2.2H_2O$ and 184 mg/L $MgSO_4.7H_2O$, and buffering with 8.5 mg/L KH_2PO_4, 21.75 mg/L K_2HPO_4 and 17.7 mg/L Na_2HPO_4. The final pH-value of the suspensions ranged from 6.6 to 6.8, while the electrical conductivity ranged from 980 to 1024 µS/cm.

To determine *width* and *length* of the *E. coli* cells, a light microscope (Olympus BX51) in phase contrast mode, with a camera (Olympus DP2) mounted on top and connected to a computer, was used to take images of cells. Per *E. coli* strain 30 different images were imported into an image processing program (DP-Soft 2) and the average cell width and cell length were determined. The equivalent spherical diameter (ESD) was determined as the geometric mean of average length and width (Rijnaarts et al., 1993).

6.2.3 Porous media

The porous media comprised of 99.1% pure quartz sand (Kristall-quartz sand, Dorsilit, Germany) with sizes ranging from 180 to 500 μm, while the median of the grain size weight distribution was 356 μm. With the median of the grain size weight, straining was excluded as a possible retention mechanism in the column set up: assuming a bacteria equivalent spherical diameter of 1.5 μm, the ratio of colloid and grain diameter was 0.004, which was well below the ratio (0.007) for which straining was observed by Bradford et al. (2007) for carboxyl latex microspheres with a diameter of 1.1 mm suspended in solutions with ionic strengths up to 31mM (the ionic strength of the solutions we used was 4.7 mmol/L only). Total porosity was determined gravimetrically to be 0.38. Prior to the experiments, the sand was rinsed sequentially with acetone, hexane and concentrated HCl, followed by repeated rinsing with de-mineralized water until the electrical conductivity was below 3 μS/cm. This was done to remove impurities.

6.2.4 Column experiments

To study the transport of the six *E. coli* strains isolated from the Kampala springs, column experiments were conducted in artificial groundwater (AGW) prepared as described above. The column consisted of a straight tube of 1.5 m transparent acrylic glass (Perspex) with an inner diameter of 9 cm, and with five sampling ports placed at 20-40 cm intervals along the tube. A stainless steel grid for supporting the sand was placed at the bottom of the tube. The column was gently filled with the clean quartz sand under saturated conditions, while the sides of the column were continuously tapped during filling, to avoid layering or trapping of air. The column was connected both at the funnel shaped effluent end and influent end with two Masterflex pumps (Console Drive Barnant Company Barrington Illinois, USA) via teflon tubes, and the pumps were adjusted to a mean fluid approach velocity of 1.16×10^{-4} m/s. Prior to a column experiment, the column was flushed for 18 hours with AGW to arrive at stable fluid conditions inside the column. Bacteria influent suspensions were prepared by washing and centrifuging at 4600 rpm for 10 minutes three times in AGW. Bacteria cell concentrations of the influent suspension were approximately 10^9 cells/mL. Experiments were conducted by applying a pulse of 0.3 PV (approximately 1.1L) of bacteria influent suspension to the column, followed by bacteria free AGW. Samples were taken at 5 distances from the column inlet except for the experiment with SL20 where samples were taken at 4 sampling distances. Samples were diluted 3 times in AGW and the optical density was measured at 254 nm using a spectrophotometer (Cecil 1021; Cecil Instruments Inc., Cambridge, England). Bacteria inactivation was assessed in all experiments by plating samples of the influent at 30 minutes intervals during the entire experiment. All plates were incubated at 37 °C for 24 hours. After each experiment, to clean the sand in the column, and to prepare for the next experiment, a pulse of 0.5 L 1.9 M HCl followed by a pulse of 0.5 L 1.5 M NaOH was flushed through the column, followed by flushing with AGW water until the electrical conductivity and pH of the effluent was equal to that of the AGW.

6.2.5 Segment sticking efficiency and bacteria fraction retained

To verify the distribution in cell affinity for collector surfaces, sticking efficiencies within column segments were computed for a fraction of cells retained in a column slice as (Lutterodt et al., 2009a, Chapter 2 of this thesis, Martin et al., 1996):

$$\alpha_i = -\frac{2}{3}\frac{d_c}{(1-\theta)\eta_0 L_i}\ln\left(\frac{M_i}{M_{i-1}}\right) \tag{6.1}$$

where α_i is the dimensionless sticking efficiency of a column slice i, d_c is the median of the grain size weight distribution (m), η_0 is the single collector contact efficiency (-), θ is the total porosity of the sand (-), and L_i is the length of the column slice i, i.e. the distance (m) between two sampling ports. M_{i-1} is the total number of cells entering slice i, obtained from the breakthrough curve determined at the upper sampling port of slice i and M_i is the total number of cells, obtained from the breakthrough curve determined at the lower sampling port of slice i using (Kretzschzmar et al., 1997):

$$M_i = q\int_0^t C(t)dt \tag{6.2}$$

where q is the volumetric flow rate (ml/min), C is the cell suspension (# cells/ml) and t is time (min). The Tufenkji-Elimelech (TE) correlation equation (Tufenkji and Elimelech, 2004a) was used to compute η_0. Thereto, 1055 kg/m³ was assumed for the bacteria density, while the Hamaker constant was estimated to be 6.5×10^{-21} J (Walker et al., 2004).

The number of retained bacteria in a quartz sand column slice, F_i, as a fraction of the total number of bacteria cells injected in the column was calculated as (Lutterodt et al., 2009b, 2011; Chapters 4 & 5):

$$F_i = \frac{M_i - M_{i-1}}{M_0} \tag{6.3}$$

where M_0 is the total number of E. coli cells in the influent suspension.

6.3 Data analysis

A power-law distribution was used to assess the degree of association between F_i and α_i and the coefficient of determination (R^2) was applied to evaluate the goodness of fit. Fitting was considered excellent when $R^2 > 0.90$, and for $0.8 \leq R^2 \leq 0.9$ fit was considered good and when $R^2 <0.80$ fit is considered weak. With these power-law equations the minimum sticking efficiencies were calculated. In order to do this, F_i was assumed to be 0.001% of the initial E. coli mass introduced into the column.

6.4 Results

6.4.1 Breakthrough curves and segment sticking efficiencies

For all the six strains, the maximum peak breakthrough concentration reduced with increasing transport distance, similar to strain SL03 (Fig. 6.2). The segment sticking efficiency, α_i (Fig. 6.3), reduced with increasing transport distance. Usually, α_i-values were highest in the first column segment, and varied between 0.198 for strain FR02 to 0.03 for strain SL20. Lowest α_i-values were determined for the most distant column slices, from a minimum of 0.002 for FR02 to 0.1 for strain FR05.

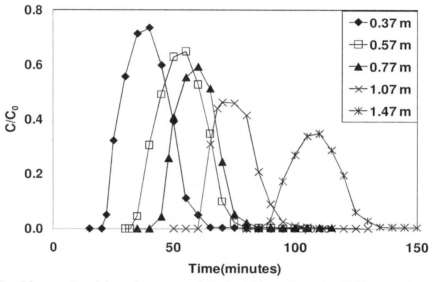

Fig. 6.2:   Breakthrough curves of Escherichia coli strain SL03 at various sampling ports (see legend) in a 1.5 m saturated quartz sand column

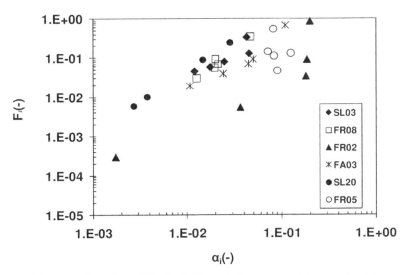

Fig. 6.3:   *Fraction of Escherichia coli input mass (F_i) retained in a column segment as a function of segment sticking efficiency (α_i)*

6.4.2 Minimum sticking efficiencies

For most of the strains, the relation between F_i and α_i could be well described with a power law distribution function (Table 6.1): coefficients of determination (R^2) ranged from 0.77 (SL03) to 0.99 (FA03). Calculated minimum sticking efficiencies, α_{min}, of the strains, were mostly in the order of 10^{-5} to 10^{-4} (Table 6.2). Strain FR05 behaved differently: for this strain, the segment sticking efficiency was invariably around 0.1, and therefore, also the minimum sticking efficiency was estimated to be 0.1. In fact, due to this relatively high α_{min} the strain was almost completely eliminated after passing the 1.5 m quartz sand column.

Table 6.1:   *Power-law distribution functions of Escherichia coli mass fraction retained (F_i) in a saturated quartz sand column segment and the segment sticking efficiencies (α_i). Extrapolated minimum sticking efficiencies (α_{min}) are given in the column on the far right*

Strain	Power law	R^2(-)	α_{min} (-)
SL03	$F_i = 8.78\alpha_i^{1.22}$	0.76	1.34×10^{-5}
FR08	$F_i = 100.53\alpha_i^{1.86}$	0.96	1.72×10^{-4}
FR02	$F_i = 1.11\alpha_i^{1.35}$	0.84	1.83×10^{-4}
FA03	$F_i = 10.29\alpha_i^{1.47}$	0.91	8.13×10^{-5}
SL20	$F_i = 68.46\alpha_i^{1.58}$	0.99	4.72×10^{-5}
FR05	$F_i = 0.08\alpha_i^{0.05}$	0.03	~0.1

Table 6.2:   *Comparison of minimum sticking efficiencies (α_{min}) obtained from this study (left two columns) and from Lutterodt et al. (2009b, 2011; Chapters 4 and 5)*

Strain	α_{min} (−)	Strain	α_{min} (−) (Lutterodt et al., 2009b)	α_{min} (−) (Lutterodt et al., 2011)
SL03	1.34×10^{-5}	UCFL-71	4.70×10^{-2}	
FR08	1.72×10^{-4}	UCFL-94	6.06×10^{-6}	1.68×10^{-7}
FR02	1.83×10^{-4}	UCFL-131	9.44×10^{-3}	6.75×10^{-8}
FA03	8.13×10^{-5}	UCFL-167	1.73×10^{-1}	
SL20	4.72×10^{-5}	UCFL-263	3.39×10^{-2}	
FR05	~ 0.1	UCFL-348	≥ 1	

6.5 Discussion

From our experiments we concluded that the segment sticking efficiency, α_i, was not a constant, but reduced with increasing transport distance. Furthermore, power law distribution functions were capable of adequately describing the relation between F_i and α_i, and minimum sticking efficiencies (α_{min}), calculated from these power law distribution functions usually ranged from 10^{-5} to 10^{-4}.

6.5.1 Attachment variation within and among bacteria strains

Our observation that the six *E. coli* strains revealed variation in their interaction with quartz sand, as witnessed by the varying segment sticking efficiencies, α_i, is consistent with observations made for *E. coli* strains isolated from different sources (Schinner et al., 2010, Bolster et al., 2009, Bolster et al., 2006, Lutterodt et al., 2011, 2009b). Such reductions in α_i with increasing transport distance have been explained in literature (Lutterodt et al., 2009a, 2009b, 2011; Chapters 2,4 and 5 of this thesis; Tufenkji and Elimelech, 2005b, 2004b; Simoni et al., 1998; Baygents et al., 1998) by heterogeneity in cell surface characteristics. Many workers have attributed higher sticking efficiencies obtained at transport distances near influent end of a column to preferential removal of cells with characteristics that promote their attachment to collector surfaces (Foppen et al., 2007a, 2007b; Bolster et al., 1999, 2000; Li et al., 2004). The low aspect ratio between the average collector grain size and average size of the strain used enabled us to rule out the possibility of straining (see Methods Section). The observed α_i variations of 1.1 log-unit within the *E. coli* strains are consistent with our previous work, despite the fact that growth conditions of the *E. coli* strains applied in the two studies differed. We can conclude from this that for *E. coli* strains isolated from different sources, though the magnitude of cell attachment differed; the transport behavior was similar with respect to intra-strain attachment variations.

6.5.2 Distributions in sticking efficiencies and minimum sticking efficiencies

The good power law type of relation between F_i and α_i for all strains (with the exception of FR05) is also consistent with results of others (Lutterodt et al., 2009b, 2011; Brown and Abramson, 2006; Tufenkji et al., 2003; Redman et al., 2001a, 2001b). The power-law distribution was used to calculate the so-called minimum sticking efficiency, α_{min}, of the strains, which is defined as the sticking efficiency belonging to a bacteria fraction of 0.001% of initial bacteria mass flowing into a column, after removal of 99.999% (5 log reduction) of the original bacteria mass has taken place. The minimum sticking efficiency practically represents the sticking efficiency of a minor fraction of bacteria cells. However, within this minor fraction of bacteria cells, the sticking efficiencies are again not constant, but they are distributed, and therefore, within this sub-fraction, the minimum sticking efficiency is the highest possible sticking efficiency (Lutterodt et al., 2011). The low values of α_{min} in the order 10^{-4} and 10^{-5} obtained from this study were compared with those obtained from Lutterodt et al. (2009b, 2011; Table 6.2). The UCFL strains were obtained from the soil of a dairy farm, where cows were grazing, and as such have not been subject to transport in aquifers. At the beginning of this study, we hypothesized that if the segment sticking efficiency reduced with transport distance, then the minimum sticking efficiency of the *E. coli* strains harvested from springs was also expected to be low to very low. Looking at Table 6.2, we conclude that, because the α_{min} of FR05 was very high (around 0.1), and because the α_{min}-values of UCFL-94 and UCFL-131 were much lower (in the order of 10^{-7}) than the values determined in this study; we had to reject our hypothesis. However, for most of the strains in our laboratory column set-ups, the segment sticking efficiency indeed reduced with transport distance. Possibly, the aquifer is heterogeneous, there are preferential flowpaths, the spring protection structure is not optimally designed and functioning or a combination of these factors was true, which all confounded the relationship between transport and minimum sticking efficiency of strains harvested from the environment.

If we eliminate our hypothesis formulated at the introductory section of this study, and if we consider strain FR05 to be an outlier, then we are left with a well defined and narrow range of minimum sticking efficiencies. We would like to argue that the importance of our work is in this set of values: for worse case scenarios in predicting pathogen transport in aquifers, sticking efficiency values of 10^{-4} to 10^{-5} should not be considered unrealistically low. Instead, we demonstrated that these values are likely characteristic for most of the *E. coli* strains that have undergone transport through an aquifer.

6.7 Conclusions

The segment sticking efficiency, α_i, of the six *E. coli* strains harvested from various springs in Kampala,Uganda, was not a constant, but reduced with increasing transport distance. Power-law distribution functions were capable of adequately describing the relation between F_i and α_i, and minimum sticking efficiencies (α_{min}), calculated from these power-law distribution functions usually ranged from 10^{-5} to 10^{-4}. For worse case scenarios in predicting pathogen transport in aquifers, sticking efficiency values of 10^{-4} to 10^{-5} should not be considered unrealistically low. Instead, in this work, we demonstrated that these values were characteristic for most of the *E. coli* strains that had undergone transport through the shallow aquifers of Kampala.

Chapter 7 Transport of *Escherichia coli* strains isolated from springs in Kampala, Uganda

This chapter is based on:
G. Lutterodt, J.W.A. Foppen and S. Uhlenbrook (2011): Transport of Escherichia coli strains harvested from springs in Kampala, Uganda. Submitted to Water Research.

Abstract

We hypothesized that the transport of *E. coli* strains harvested from termination points of flow lines in aquifers (e.g. springs) could possibly be characterized by a rather homogeneous set of *E. coli* surface characteristics and transport parameters. In order to test our hypothesis, we sampled 77 springs throughout the Lubigi catchment, which is part of Kampala, the capital of Uganda. Of the spring water samples, besides chemical and physical parameters, also thermotolerant coliform concentrations were determined. Furthermore, *E. coli* strains were harvested, and cell properties (serotype, hydrophobicity, motility, zeta potential, cell aggregation, and cell size) were determined. Then, of 40 randomly selected *E. coli* strains transport experiments in saturated quartz columns of 7 cm height were carried out to determine transport characteristics of the strains. Using a two-site non-equilibrium sorption model, transport was modeled with HYDRUS 1-D, and fitted with measured breakthrough data. The results demonstrated fecal contamination of the springs with high thermotolerant coliform concentrations and also high concentrations of chloride and nitrate. The transport of the *E. coli* strains was remarkable similar: some 82 % of the strains had a maximum relative breakthrough concentration between 0.5 and 1, while some 75% of the 40 strains had similar attachment efficiency values in the order of 10^{-3} and 10^{-4}. We attributed this to the way in which the strains were harvested: from springs, and therefore at the termination points of flowlines. Such strains may indeed possess certain cell characteristics that might have influenced their selective transport in the subsurface giving rise to their similar transport characteristics in our columns. There was however no statistically significant correlation between measured cell properties (serotype, zeta potential, motility, cell size, cell aggregation, and hydrophobicity) and transport parameters (f, ω and k_s and $(C/C_0)_{max}$).

Furthermore, 58% of the strains we tested were of the O21:H7 serotype. This suggests that this specific serotype may possess certain characteristics that allows its preferential transport through the shallow aquifers, in the Kampala area. Our work demonstrated that in order to assess transport characteristics of *Escherichia coli* in real world surroundings, it seems to be better to harvest bacteria from the aquifer itself rather than from pollution sources, like fecal sludge, solid waste piles or grey water disposal sites.

7.1 Introduction

Groundwater is an immensely important resource, as it provides more than one-third of the world's drinking water (Morris et al., 2003). An important characteristic that has often been associated with groundwater is the assumption that, generally, the resource is free of pathogenic microorganisms (Bhattacharjee et al., 2002). In spite of the cleanliness of groundwater, many water borne disease outbreaks are caused by the consumption of groundwater contaminated by pathogenic bacteria and protozoa (Close et al., 2006, Powell et al., 2003, Macler and Merkle, 2000). According to the WHO, an estimated one billion people lack access to an improved water supply and two million deaths are attributable to unsafe drinking water, sanitation and hygiene, with many countries still reporting cholera to the WHO (WHO, 2004). The use of shallow groundwater for drinking and domestic purposes is a common feature in many developing countries (Kulabako et al., 2007). In Kampala, the capital of Uganda, protected springs within the shallow aquifer are a major source of water supply (Kulabako et al., 2007). The springs are susceptible to pollution related to anthropogenic activities, even when protected (Kulabako et al., 2008). Previous studies undertaken on the protected springs in the area have indicated widespread faecal contamination (Kulabako et al, 2007; Howard et al., 2003; Byamukama et al., 2000).

Due to the importance of *E. coli* as a fecal indicator organism, considerable attention has been given to understanding their transport and fate in saturated porous media (e.g. Foppen et al., 2006, Schinner et al., 2010, Bolster et al., 2010). In most of these studies, bacteria strains isolated from different sources, for example from zoo animals (Foppen et al., 2010), a swine lagoon (Bolster et al., 2010), a dairy cow manure and sewerage (Haznedaroglu et al., 2008), or a soil of a pasture used for cattle grazing (Foppen et al., 2010; Lutterodt et al., 2009a, 2009b; Yang et al., 2008) were used. To our knowledge, there is no study reported in the literature that focuses on the transport of *E. coli* strains isolated *after* they have undergone transport in an aquifer. These strains may start transport among a wide variety of microorganisms that have infiltrated from the surface into the groundwater system, and, after considerable transport has taken place, they may end up as one of the few microorganisms remaining in the system.

Our objectives were two-fold: 1. To present an assessment of the chemical, physical and bacteriological status of the springs in the Kampala area of Uganda, and 2. to characterize the transport of *Escherichia coli* strains isolated from these springs, when considerable transport through the aquifer has already taken place. The underlying hypothesis was that transport by such a group of *E. coli* strains could possibly be characterized by a rather homogeneous set of surface characteristics and transport parameters. To determine transport parameters, we employed saturated laboratory column experiments with saturated pure quartz sand.

7.2   Materials and methods

7.2.1 Study area

The Lubigi catchment (Fig. 7.1) in the Kampala area of Uganda is underlain by a variety of meta-sedimentary rocks and weathering has produced a pronounced topography. The area is low lying with a high water table (<1.5 m) in the weathered regolith (Kulabako et al.,2007) and site investigations showed the presence of preferential flow paths, although it is not known how far these extend (Howard et al., 2003). Previous studies have shown that the top of the regolith is composed of fine material, with increasingly coarser (sandy clay) material found at depth with productive aquifers commonly associated with these layers (Howard et al., 2003; Kulabako et al., 2007). Recharge tends to occur during two distinct wet seasons, although Kampala experiences rainfall throughout the year due to its proximity to Lake Victoria (Howard et al., 2003, Nyenje et al., 2010)

7.2.2 Sample collection and analyses

To assess the state of pollution of the springs in the Kampala area, a total of 77 springs were sampled. The locations of the springs are indicated on Fig. 7.1.

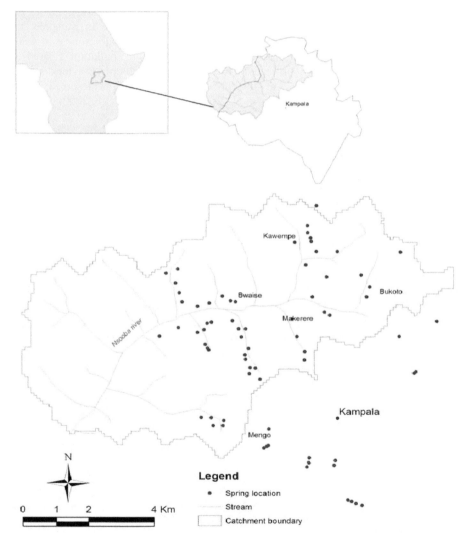

Fig. 7.1:   *Map showing spring sampling locations in the study area (Lubigi catchment) in Kampala, Uganda, insert map indicates location within East Africa.*

7.2.3 Analyses of Chemo-physical parameters

Temperature and EC were determined using a conductivity meter Cond340i (WTW GmbH, Weilheim Germany) calibrated at 25°C and pH was determined using a pH meter pH340i (WTW GmbH, Weilheim Germany). To analyze nitrate and chloride, as chemical indicators of contamination, 250 ml of spring water was collected in sterile polypropylene bottles and by means of a syringe 25 ml of sample was filtered through 0.45 µm cellulose acetate filter into scintillation vials. Samples were then stored in a cool box and transported to the Environmental Engineering and Public Health (EEPH) laboratory of the University of Makerere and stored at -20°C until they were transported to the UNESCO-IHE laboratory in Delft, The Netherlands, where nitrate and chloride were analyzed with an ICS-1000 AS 14A Column (Dionex, Benelux BV).

7.2.4 Microbiological analyses and E. coli isolation

Total coliform (TC) in spring waters were analyzed by filter pressing 100 ml of spring water and the filter paper was placed on a Chromocult™ agar (Merck, Whitehouse Station, NJ) plate. Agar plates were then transported to the EEPH laboratory of Makerere University and incubated at 37°C for 24 hours, respectively. The number of thermotolerant coliforms on plates was then counted, and the purple colored colonies on the Chromocult agar plates allowed for the detection and isolation of E. coli from other types of bacteria species growing on the agar plates. To do so, a sterile toothpick was used to pick a single colony of E. coli from the agar plates and inoculated into 5 ml of Nutrient Broth followed by incubation at 37°C for 24 hours. Then, 1-2 ml of freshly grown E. coli cells were inoculated into sterile vials (Microbank™-Dry, PRO-LAB DIAGNOSTICS, Toronto, Canada) containing porous beads saturated with cryopreservative (cryovials), which serve as carriers to support microorganisms. The vials were then stored at -20°C and transported to the UNESCO-IHE laboratory in Delft, the Netherlands.

7.2.5 Bacteria growth and cell characterization

To conduct experiments, a sterile forceps was used to pick a bead from the cryovial, placed into 25 ml of Nutrient Broth and then incubated for 24 hours at 37 °C while shaking at 150 rpm on an orbital shaker to obtain a cell concentration of ~10^9 cells/ml. Bacteria were washed and centrifuged (4600 rpm) three times in Artificial Ground Water (AGW), which was prepared by dissolving 526 mg/L $CaCl_2.2H_2O$ and 184 mg/L $MgSO_4.7H_2O$, and buffering with 8.5 mg/L KH_2PO_4, 21.75 mg/L K_2HPO_4 and 17.7 mg/L Na_2HPO_4. The final pH-value of the suspensions ranged from 6.6 to 6.8, while the electrical conductivity ranged from 980 to 1024 µS/cm

To determine *motility*, a 2 mL fresh culture was centrifuged (14000 xg) and washed three times in AGW, and by means of a sterile toothpick, cells were picked from the remaining pellet in the test tube and inoculated at the centre of petri-dishes containing 0.35% Chromocult agar. The plates were incubated at 37 °C for 24 hours after which growth and diameter of migration was measured as motility (Ulett et al., 2006).

Hydrophobicity was determined with the Microbial Adhesion To Hydrocarbons (MATH) method (Pembrey et al., 1999; Walker et al., 2005), where percentage partitioning of cells into dodecane was measured as cell hydrophobicity. Thereto, 4 mL of bacteria suspension of known optical density and 1 mL of dodecane were vigorously mixed in a test tube for two minutes and left to stand for 15 minutes to allow phase separation. Then, the optical density of the aqueous phase was determined, and the percentage of cells partitioned into the hydrophobic substance was reported as percentage hydrophobicity. All optical densities were measured at an absorbance of 254 nm.

To determine *cell aggregation*, 15 ml of freshly grown bacteria were centrifuged (14000 xg) and washed three times in AGW, and, then, allowed to stand for 180 minutes at a temperature of 4 °C. A sample of 1 mL 1 cm below the surface of the suspension was obtained, immediately and 180 minutes after washing. The optical density of the samples was measured at 254 nm, and the auto-aggregation was determined as the ratio of the final over the initial optical density (in %).

To determine the *zeta potential*, a zeta-meter similar to the one made by Neihof (1969) was used. Movement of bacteria was visible on a video screen attached to a camera mounted on top of a light microscope (Olympus EHT) in phase contrast mode (Foppen et al., 2007). Bacteria mobility values were obtained from measurements on at least 50 bacteria cells. Velocity measurements were used to calculate the zeta potential with the Smoluchowksi equation.

To measure the *average equivalent spherical diameter* of the cells, a light microscope (Olympus BX51) in phase contrast mode, with a camera (Olympus DP2) mounted on top and connected to a computer, was used to take images of cells. At least 50 images were imported into an image processing program (DP-Soft 2) and the average cell width and cell length were measured. The equivalent spherical diameter was determined as the geometric mean of average length and width (Rijnaarts et al., 1993).

7.2.6 Serotyping

Pure cultures of *E. coli* were grown on Chromocult agar and sent to the Dutch National Institute of Public Health and Environmental Hygiene (RIVM), Utrecht, The Netherlands, where serotyping was performed according to standard procedures. For O typing, the strains were tested against O1 antiserum until O181 antiserum using the classical approach (Guinée et al., 1972). For H typing, the strains were incubated 30°C and test performed in small glass tubes against H antisera H1 until H56. Details of the method are described in Ewing (1986).

7.2.7 Porous media and transport experiments

The porous media comprised of 99.1% pure quartz sand (Kristall-quartz sand, Dorsilit, Germany) with sizes ranging from 180 to 500 μm, while the median of the grain size weight distribution was 356 μm. With this grain size, we excluded straining as a possible retention mechanism in our column: assuming a bacteria equivalent spherical diameter of 1.5 μm, the ratio of colloid and grain diameter was 0.004, which was well below the ratio (0.007) for which straining was observed by Bradford et al. (2007) for carboxyl latex microspheres with a diameter of 1.1 mm suspended in solutions with ionic strengths up to 31mM (the ionic strength of the solutions we used was 4.7 mmol/L only). Total porosity was determined gravimetrically to be 0.39. The quartz sand was rinsed sequentially with acetone, hexane and concentrated HCl, followed by repeated rinsing with de-mineralized water until the electrical conductivity was close to zero (Li et al., 2004).

To assess the transport characteristics and attachment variations among *E. coli* strains isolated *after* they had undergone transport in an aquifer, forty of the seventy-one *E. coli* strains were randomly selected for column experiments. These experiments were conducted in borosilicate glass columns with an inner diameter of 2.5 cm (Omnifit, Cambridge, U.K.) with polyethylene frits (25 μm pore diameter) and one adjustable endpiece. The column was packed wet with the quartz sand with vibration to minimize any layering or air entrapment. The column sediment length was 7 cm. All column experiments were conducted in artificial groundwater (AGW) at a velocity of 0.25 PV per minute (fluid approach velocity = 10^{-4} m/s). Prior to each experiment, and in order to remove retained cells of the previous experiment, the column was rinsed with 1 PV of 1.9 M HCl, immediately followed by a pulse of 1.5 M NaOH to restore pH. The column was then equilibrated by flushing for 50-60 pore volumes with AGW to restore pH and EC. To conduct bacteria transport experiments, a suspension of *E. coli* with a concentration of ~10^9 cells/mL was flushed through the column for 4 minutes (approximately equal to one pore volume) followed by a flush of *E. coli*-free AGW. The *E. coli* concentration was determined using optical density measurements (at 254 nm) with a 1 cm flow-through quartz cuvette and a spectrophotometer (Cecil 1021, Cecil Instruments Inc., Cambridge, England). Cell numbers were deduced after calibration with plate counts on Chromocult™ agar (Merck, Whitehouse Station, NJ). To check whether the flush with HCl followed by NaOH had indeed removed all bacterial cells, at the beginning of each experiment, effluent samples were plated in triplicate. All plates of all experiments were negative, indicating that, after the previous experiment, all viable bacterial cells had indeed been removed from the column.

7.2.8 Transport model

The one dimensional (macroscopic) mass balance equation for mobile bacteria suspended in the aqueous phase excluding bacteria growth and decay is normally expressed as (Corapcioglu and Haridas, 1984, 1985.; Foppen et al., 2007)

$$\frac{\partial c}{\partial t} = D\frac{\partial^2 C}{\partial x^2} - v\frac{\partial C}{\partial x} - \frac{\rho_{bulk}}{\theta}\frac{\partial S}{\partial t} \tag{7.1}$$

Where C is the mass concentration of suspended bacteria in the aqueous phase (# of cells/ml), t is time (min), D is the hydrodynamic dispersion coefficient (cm^2/min), v is the pore water flow velocity (cm/min), S is total retained bacteria concentration (#cells/g) ρ_{bulk} is the bulk density (g/ml), x is the distance traveled (cm), and θ is the volume occupied by the fluid per total volume medium (-). We applied the two-site sorption model (Cameron and Klute, 1977) in this study. The model describes the interaction of mass between the aqeous phase and the solid phase by a first order kinetic reaction together with instantaneous equilibrium sorption. The model assumes that, interaction can be divided into two fractions

$$S = S_e + S_k \tag{7.2}$$

where S_e is the mass adsorbed at equilibrium sites (type-1 sites), S_k is the mass adsorbed at kinetically controlled sites (type-2 sites). The relation between the total adsorbed mass and the mass adsorbed at type-1 and type-2 sorption sites respectively are

$$S_e = fS \tag{7.3}$$
$$S_e = fk_sC \tag{7.4}$$
$$S_k = (1-f)S \tag{7.5}$$

where f represents the fraction of exchange sites assumed to be in equilibrium with the solution phase (-), and k_s an equilibrium sorption coefficient (ml/g). The time rate of change of mass at type-1 sites is given by

$$\frac{\partial S_e}{\partial t} = f \frac{\partial S}{\partial t}$$ (7.6)

and for type-2 sites, the time rate of change of mass is given by

$$\frac{\partial S_k}{\partial t} = \omega \left[(1-f)k_s C - S_k \right]$$ (7.7)

where ω is a first order rate constant (min^{-1}). The parameters f, ω and k_s were obtained by fitting the model to measured bacteria breakthrough curves in HYDRUS-1D (Šimůnek et al., 2009). Then, the bacteria attachment efficiency, α (-), was computed using the relation

$$\frac{3(1-\theta)}{2d_c} \eta v \alpha = \omega \left[(1-f)k_s \right] \frac{\theta}{\rho_{bulk}}$$ (7.8)

Where η (-) is the single collector contact efficiency, and d_c is mean grain diameter (cm). We determined η with the Tufenkji-Elimelech correlation equation (Tufenkji and Elimelech, 2004a) assuming a particle density = 1055 kg/m^3, fluid density = 1000 kg/m^3, fluid viscosity = 1.005×10^3 kg/m s, temperature = 278 K, and a Hamaker constant = 6.5×10^{-21} J (Walker et al., 2004).

7.3 Statistical analysis

Correlation between cell characteristics and transport parameters obtained from HYDRUS modeling were determined with the Pearson Correlation Test, also Fisher's least significant difference (LSD) was applied using analysis of variance (ANOVA) to compare the mean values of transport characteristics and also measured cell properties of identified serogroups. All statistical analyses were performed using SPSS 14 (SPSS Inc. Chicago).

7.4 Results

7.4.1 Chemo-physical and bacteriological characterization of the springs in Kampala

Temperature values for all springs were 24.0 ± 0.54 °C, while pH ranged between 5.14 ± 0.36. Nitrate and chloride concentrations of the samples were high and ranged from 4.7 to 135.8 mg/l and from 13.9 to 151.2 mg/l, respectively (Fig. 7.2). Based on the chemistry, we concluded that the springs samples were polluted, likely resulting from the infiltration of waste water. With the exception of two springs (K3, NaK-1), all springs we had sampled contained thermotolerant coliforms with concentrations ranging from 10^3 to 1.9×10^5 cells/100 ml. The observed good relation (R^2 =0.87, p = 0.00) between nitrate and chloride content (Fig. 7.2) indicated their possible release from the same source. The lack of direct correlation between nitrate or chloride and total coliforms suggested the possibility of multiple sources of coliforms leaked into the groundwater systems of the area.

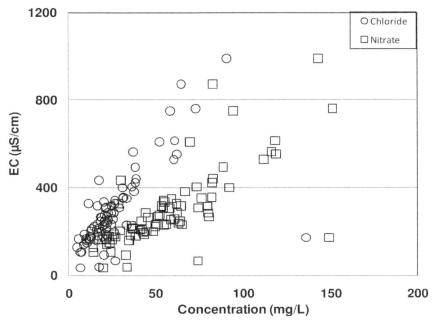

Fig. 7.2:   *Relation between EC and nitrate and chloride concentrations*

7.4.2 Bacteria characterization (serotyping and cell characteristics)

The *E. coli* serotyping showed four different O serogroups (O14, O21, O91, and O108), while 16 of the strains were O-untypable and six strains were not typed. The results indicated that 58 % of the serotyped *E. coli* strains exhibited the same serotype (O21 H7), while the remaining identified serotypes were O108:H45, O14:H15, O14:H4, and O91:H28. Identified H types with untypable O-serogroups were H31, H8, and H6. For some of the strains, both the O serogroup and H type were untypable (O-nt: H?; see Table 7.1).

Cell characteristics (motility, hydrophobicity, cell aggregation, zeta potential and average cell size) of 40 randomly selected *E. coli* strains were determined. Cell motility measured as the growth and diameter of migration on agar plates ranged between 0.2 cm (SL43 and FA01) to 9 cm (SL21 and H31; Table 7.2). The hydrophobicity varied from less than 1% (FR05) to 44.6% (SL20), and auto-aggregation ranged from no auto-aggregation (SL41) to 64% (SL14), respectively. Average cell surface charge ranged from -12.4 mV for FR04 to -31.2 mV for IN01, whereas the average size ranged from 1.76 µm for SL08 to 3.22 µm for FR04. Apart from the measured average cell size, the standard deviations from the mean values of all measured cell characteristics were high. From these results, we concluded that the cell properties varied considerably among the strains.

Table 7.1: Chemo-physical and bacteriological characteristics of the springs

Spring ID	Location North	Location East	pH	EC (µS/cm)	Temp (°C)	TC (cells/100mL)	Cl (mg/L)	NO$_3$ (mg/L)	Strain isolated from spring	Strain serotype
K01	00°20.918'	32°34.787'	4.89	238	23.0	1×10^7	22.2	63.19	SL01-SL03	O21:H7
K02	00°20.865'	32°34.868'	6.02	267	25.0	1×10^5	22.1	51.37	SL04-SL06	O21:H7
K03	00°22.014'	32°34.990'	5.3	127	25.0	6×10^5	4.7	13.91	NSI	
K04	00°22.013'	32°34.659'	4.94	148	25.0	8×10^7	10.6	23.56	SL23	O-nt :H?
K05	00°21.592'	32°35.375'	5.09	105.4	25.0	1×10^5	7.4	24.33	SL07-SL08	O21:H7
K06	00°22.006'	32°36.006'	5.01	159	23.0	1×10^5	10.3	34.91	SL09-SL11	O21:H7
K07	00°21.379'	32°35.513'	4.16	187	23.0	1×10^5	9.1	26.05	SL16-SL17 SL12-SL13	O21:H7 O21:H7
									SL18	O21:H7
K08	00°21.195'	32°35.464'	4.8	196	23.5	2×10^5	16.5	42.82	SL14	O21:H7
									SL25	O-nt :H?
K09	00°21.296'	32°35.473'	5.12	214.6	23.6	1×10^5	17.5	36.90	SL28	O14:H4
K10	00°21.108'	32°33.352'	4.4	553.8	23.8	1×10^5	62.2	119.09	SL15	O21:H7
									SL24	O-nt:H 3)
K11	00°21.125'	32°33.273'	4.85	528.3	23.6	2×10^7	60.5	112.02	SL29	O-nt :H8
K12	00°21.210'	32°33.140'	4.8	440.4	24.4	1×10^5	38.9	82.82	SL27	O-nt :H8
K13	00°20.737'	32°32.965'	4.84	245.6	23.4	3×10^5	20.7	63.98	SL26	O-nt :H?
K14	00°20.616'	32°33.403'	4.87	285.3	23	1×10^5	26.2	62.36	SL21	O-nt :H31
K15	00°20.618'	32°33.515'	5.18	163.2	23.5	2×10^5	12.0	25.19	SL19	O-nt :H31
K16	00°20.549'	32°32.743'	5.06	308	24.6	4×10^5	29.1	74.28	NSI	
K17	00°20.600'	32°32.849'	4.9	177.6	23.8	1×10^5	16.9	38.68	SL22	O-nt :H6

Table 7.1: Chemo-physical and bacteriological characteristics of the springs (Contd.)

Spring ID	Location		pH	EC (µS/cm)	Temp (°C)	TC (cells/100mL)	Cl⁻ (mg/L)	NO₃ (mg/L)	Strain isolated from spring	Strain serotype
	North	East								
K18	00°20.718'	32°32.899'	4.74	199.5	24.0	TNTC*	26.2	48.71	SL20	O-nt:H8
K19	00°20.735	32°32.964'	4.95	275.1	23.6	7×10^6	20.8	52.22	NSI	
K20	00°20.639'	32°32.436'	4.96	327	24.1	4×10^6	26.7	61.93	SL31	O21:H7
K21	00°20.476'	32°32.129'	4.71	254	23.7	5×10^6	22.9	58.63	NSI	
K22	00°20.330'	32°32.869'	5.10	285	24.3	7×10^6	22.6	44.40	NSI	
K23	00°20.264'	32°32.008'	5.16	230	23.9	9×10^6	33.5	54.53	NSI	
K24	00°20.228'	32°32.922'	5.06	306	23.9	3×10^6	21.3	58.01	NSI	
K25	00°20.763'	32°33.404'	4.83	382	24.7	1×10^6	37.6	66.76	NSI	
K26	00°20.467'	32°33.452'	4.94	341	24.4	3×10^6	26.6	54.39	NSI	
K27	00°20.259'	32°33.564'	4.94	258	24.0	2×10^6	25.7	60.61	NSI	
K28	00°20.140'	32°33.509'	4.95	614	24.9	2×10^6	61.0	118.45	SL33	O14:H15
K29	00°20.062'	32°33.519'	5.25	761	24.3	TNTC	72.9	151.18	NSI	
K30	00°19.912'	32°33.504'	5.40	171.2	24.3	2×10^6	135.8	148.94	NSI	
K31	00°19.900'	32°33.677'	5.33	37.2	24.0	3×10^6	17.6	33.72	SL30	O21:H7
K32	00°19.800'	32°33.581'	6.03	92.7	24.4	2×10^6	20.3	32.90	NSI	
K33	00°19.700'	32°33.751'	5.02	65.7	24.4	2×10^6	27.0	74.18	NSI	
K34	00°18.954'	32°33.165'	4.95	33.2	23.4	1×10^6	6.6	19.85	FR04-FR05	O21:H7
									FR07-FR08	O21:H7
									FR09	O-nt:H31

Table 7.1: Chemo-physical and bacteriological characteristics of the springs (contd.)

Spring ID	Location North	Location East	pH	EC (µS/cm)	Temp (°C)	TC (cells/100mL)	Cl (mg/L)	NO$_3$ (mg/L)	Strain isolated from spring	Strain serotype
K35	00°19.005'	32°32.964'	4.80	232	23.7	1×10^1	24.6	56.28	FR01	O-nt:H31
									FR02	O21:H7
									FR06 & FR10	O21:H7
									FR11	O-nt:H31
K36	00°19.010'	32°32.806'	5.24	139.1	23.6	2×10^1	8.6	22.01	FA01	O21:H7
									FA03	O108:H45
									FA04-FA06	No typing
K37	00°18.862'	32°33.001'	4.98	175.9	23.6	2×10^1	12.1	23.07	FA02	O108:H45
									FA05,FA07,FA08	No typing
K38	00°18.862'	32°33.154'	5.20	315	23.8	1×10^1	24.8	79.81	FR12-FR14	No typing
NaK-1	00°20.477'	32°35.990'	5.50	103.8	24.4	0×10^0	6.7	14.21	NSI	
NaK-2	00°20.756'	32°36.590'	5.07	400	24.4	1×10^1	31.1	92.05	IN01	O-nt:H8
									IN06	O21:H7
NaK-3	00°19.840'	32°36.261'	5.88	164.5	24.9	2×10^1	10.1	21.68	IN03	O21:H7
									IN05	O-nt:H8
									IN08	No typing
NaK-4	00°19.812'	32°36.231'	5.40	174	24.7	9×10^1	12.7	27.18	IN02	O-nt:H8
K43	00°21.560'	32°34.819'	4.91	284	23.9	8×10^1	24.4	80.01	IN04 & IN07 NSI	O21:H7
K44	00°20.793'	32°34.270'	4.98	352	24.6	6×10^1	31.3	76.36	NSI	
K45	00°20.473'	32°34.346'	5.30	422	23.9	6×10^1	38.4	82.48	SL37-SL38	O21:H7

Table 7.1: Chemo-physical and bacteriological characteristics of the springs (contd.)

Spring ID	Location		pH	EC	Temp	TC	Cl⁻	NO₃	Strain isolated	Strain
	North	East		(µS/cm)	(°C)	(cells/100mL)	(mg/L)	(mg/L)	from spring	serotype
K46	00°21.701'	32°32.429'	5.02	166	23.3	1×10^2	5.7	12.18	SL39	O21:H7
K47	00°21.626'	32°32.233'	4.98	162	23.5	3×10^3	21.4	19.39	SL40	O21:H7
K48	00°21.446'	32°32.389'	4.95	158	23.7	6×10^4	15.8	34.68	NSI	
K51	00°21.024'	32°32.745'	4.91	211	23.5	1×10^4	22.0	42.02	SL41	O21:H7
K52	00°21.085'	32°32.934'	5.21	311	23.8	1×10^5	21.7	54.34	NSI	
K53	00°22.846'	32°34.655'	5.24	403	24.2	1×10^2	36.5	73.47	SL42	O21:H7
K54	00°22.483'	32°34.515'	5.25	316	24.3	1×10^3	16.6	53.48	NSI	
K55	00°22.351'	32°34.519'	5.48	266	24.5	1×10^3	18.5	80.35	SL43	O21:H7
K56	00°22.259'	32°34.567'	5.24	317	24.0	2×10^4	24.8	64.34	NSI	
K57	00°22.197'	32°34.578'	5.14	212	24.5	5×10^3	13.4	37.21	SL44	O21:H7
K58	00°22.180'	32°34.309'	5.14	232	24.3	5×10^3	16.6	50.26	NSI	
K59	00°21.777'	32°34.489'	5.50	432	24.8	2×10^2	17.6	30.23	SL45-SL46	O91:H28
K60	00°21.198'	32°34.596'	5.00	225	23.9	2×10^3	19.6	36.03	SL34	O21:H7
									SL35	O14:H15
K61	00°18.806'	32°33.893'	5.20	563	23.9	3×10^3	37.1	116.38	NSI	
K62	00°18.505'	32°33.886'	5.30	493	23.9	TNTC	38.4	88.67	NSI	
K63	00°18.496'	32°33.872'	5.15	335	23.6	1×10^3	20.3	55.13	NSI	
K64	00°18.488'	32°33.859'	5.40	265	23.4	1×10^4	19.0	45.77	NSI	
K65	00°18.460'	32°33.815'	5.20	246	23.3	1×10^4	18.2	41.81	NSI	
K66	00°18.285'	32°34.548'	5.73	750	24.6	2×10^3	58.3	94.54	NSI	

Table 7.1: Chemo-physical and bacteriological characteristics of the springs (contd.)

Spring ID	Location North	East	pH	EC (µS/cm)	Temp (°C)	TC (cells/100mL)	Cl (mg/L)	NO$_3$ (mg/L)	Strain isolated from spring	Strain serotype
K67	00°18.200'	32°34.541'	5.30	988	24.2	TNTC	90.8	142.82	SL36	O14:H15
K68	00°18.186'	32°34.545'	6.01	874	23.9	1×10^3	64.7	82.93	NSI	
K69	00°18.113'	32°34.511'	6.50	609	25.1	TNTC	52.3	69.69	NSI	
K70	00°20.050'	32°34.468'	5.69	327	23.9	4×10^2	11.3	28.65	NSI	
K71	00°20.200'	32°34.475'	5.20	187	23.7	1×10^1	21.0	43.33	NSI	
K72	00°18.239'	32°34.962'	5.20	221	23.3	2×10^1	14.7	49.70	NSI	
K73	00°18.143'	32°34.945'	5.35	350	23.6	1×10^1	33.4	59.05	NSI	
K74	00°17.513'	32°35.166'	4.98	356	23.9	TNTC	31.5	82.10	NSI	
K75	00°17.488'	32°35.224'	4.83	231	23.9	2×10^2	20.9	64.84	NSI	
K76	00°17.450'	32°35.298'	5.18	203	24.3	9×10^1	12.5	29.80	NSI	
K77	00°17.417'	32°35.398'	5.12	199	23.1	6×10^2	18.8	44.72	NSI	

7.4.3 Transport experiments and HYDRUS modeling

To characterize the transport of the *E. coli* strains isolated from springs, 40 randomly selected strains were flushed through saturated quartz columns of 7 cm height. Maximum peak relative breakthrough concentrations, $(C/C_0)_{max}$, ranged from 0.06 to 0.89 representing a 1 log unit variation, with 33 out of the 40 strains tested (82%) having a $(C/C_0)_{max}$ value greater than 0.5 (Table 7.2; Fig.7. 3).

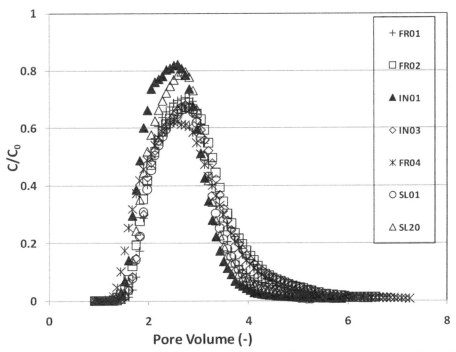

Fig. 7.3:   Examples of a number of breakthrough curves of E. coli strains with high $(C/C_0)_{max}$

Table 7.2:   *Measured properties of Escherichia coli strains (left part) and parameters obtained by fitting breakthrough curves in the HYDRUS-1D model (right part)*

	Measured *E. coli* strain property							Parameters obtained from modeling			
Strain	Serotype	Motility	Hydroph.	Cell	Zeta	Av.	$(C/C_0)_{max}$	α	f	ω	k_s
		(cm)	(%)	agg.	pot.	size	(-)	(-)	(-)	(T^{-1})	$(M^{-1}L^3)$
				(%)	(-mV)	(µm)					
FR01	O nt H31	5.50	1.89	1.44	23.21	2.41	0.71	1.41E-07	7.91E-02	1.85E-06	2.31E+00
FR02	O21 H7	7.45	4.04	47.0	27.53	2.26	0.67	1.05E-02	1.19E-01	1.68E+00	1.92E-01
FR03	O21 H7	6.90	1.33	15.6	27.25	2.45	0.38	2.00E-03	2.15E-02	9.07E-03	6.36E+00
FR04	O21 H7	5.60	6.20	1.98	12.39	3.22	0.62	6.61E-03	1.90E-04	1.28E+00	1.72E-01
FR05	O21 H7	6.25	0.21	2.77	18.96	2.12	0.64	7.04E-03	1.76E-01	1.23E+00	1.84E-01
FR06	O21 H7	8.45	7.99	15.2	19.79	2.47	0.48	1.13E-03	2.87E-01	8.89E-02	5.05E-01
FR07	O21 H7	7.25	29.86	21.2	15.40	2.46	0.61	5.11E-03	2.23E-01	9.17E-01	2.03E-01
FR08	O21 H7	6.95	8.78	54.3	13.89	1.83	0.89	2.28E-03	6.09E-01	7.84E-01	1.87E-01
IN01	O nt H8	7.65	6.88	0.70	31.18	2.44	0.81	1.14E-02	1.31E-02	2.64E+00	1.25E-01
IN02	O nt H8	6.75	3.43	1.52	26.53	1.92	0.74	7.23E-03	6.91E-02	1.40E+00	1.42E-01
IN03	O21 H7	0.35	0.59	1.50	25.74	1.85	0.68	1.11E-02	1.15E-01	1.65E+00	1.91E-01
IN05	O nt H8	8.15	19.40	2.40	25.79	2.39	0.71	1.17E-02	2.00E-05	2.01E+00	1.65E-01
IN06	O21 H7	5.95	7.25	6.16	30.65	2.12	0.71	1.07E-02	5.18E-02	1.96E+00	1.54E-01
IN07	O21 H7	6.15	13.40	0.95	21.10	2.16	0.63	2.25E-04	5.17E-01	3.91E-02	3.21E-01
SL01	O21 H7	7.30	7.44	1.85	20.60	2.17	0.67	8.73E-03	4.21E-01	2.31E+00	1.75E-01
SL02	O21 H7	6.85	11.58	28.2	13.89	2.57	0.33	2.46E-03	6.74E-02	3.44E-02	2.22E+00
SL03	O21 H7	6.05	12.69	1.98	17.50	2.23	0.68	3.49E-04	6.28E-01	1.14E-01	2.24E-01
SL04	O21 H7	6.70	13.77	26.1	18.96	2.23	0.15	5.33E-03	4.57E-02	3.50E-02	4.31E+00
SL08	O21 H7	2.10	7.18	0.73	26.68	1.76	0.66	1.47E-06	9.98E-01	8.98E-02	1.76E-01
SL12	O21 H7	4.00	11.72	1.50	25.25	2.17	0.67	1.00E-04	5.14E-01	2.04E-02	2.71E-01
SL14	O21 H7	8.60	12.17	63.8	21.30	2.38	0.18	3.62E-03	1.13E-02	6.44E-03	1.58E+01
SL16	O21 H7	7.10	13.97	2.05	18.11	2.39	0.32	2.34E-03	2.51E-02	1.33E-02	5.03E+00
SL20	O nt H8	5.65	44.63	2.60	21.46	2.34	0.80	1.60E-07	8.70E-01	2.13E-04	1.60E-01
SL21	O nt H31	9.00	5.11	0.39	19.60	2.37	0.65	6.07E-03	3.38E-01	9.90E-01	2.58E-01
SL22	O nt H6	4.75	11.40	0.00	17.73	2.17	0.57	2.99E-03	4.46E-01	5.07E-01	2.84E-01
SL28	O14 H4	6.25	4.22	3.91	18.28	1.84	0.70	5.26E-03	3.52E-01	1.00E+00	2.04E-01
SL30	O21 H7	5.30	9.50	1.70	12.76	2.30	0.06	6.85E-03	6.10E-03	8.80E-03	2.15E+01
SL31	O21 H7	7.05	6.21	54.5	19.00	2.44	0.72	1.49E-03	6.29E-01	9.03E-01	1.26E-01
SL33	O14 H15	6.65	6.92	0.45	20.50	2.39	0.71	4.55E-04	7.40E-01	2.65E-01	1.85E-01
SL37	O21 H7	2.45	7.41	44.1	20.08	1.82	0.67	6.75E-04	9.51E-01	1.72E+00	2.00E-01
SL40	O21 H7	4.05	25.82	0.96	21.17	2.34	0.74	1.72E-03	6.86E-01	8.63E-01	1.75E-01
SL41	O21 H7	0.50	10.74	0.00	18.64	2.38	0.69	4.68E-03	3.84E-01	1.00E+00	2.11E-01
SL42	O21 H7	6.70	22.71	36.1	16.20	2.18	0.66	8.55E-04	2.98E-01	7.25E-02	4.50E-01
SL43	O21 H7	0.20	2.97	4.17	19.65	1.98	0.69	1.80E-03	7.60E-01	1.00E+00	1.94E-01
SL44	O21 H7	1.30	5.45	0.68	14.30	2.32	0.67	5.76E-03	3.60E-01	1.15E+00	2.16E-01
SL45	O91 H28	1.00	27.30	3.00	26.79	1.86	0.80	1.89E-07	1.00E+00	3.31E-01	1.44E-01
SL46	O91 H28	8.70	16.53	3.35	14.74	2.45	0.54	7.87E-04	5.09E-01	1.12E-01	4.05E-01
FA01	O21 H7	0.20	11.88	0.38	12.75	2.19	0.61	5.39E-04	2.52E-01	2.94E-02	6.62E-01
FA02	O108 H45	6.15	19.36	0.85	16.12	2.14	0.59	4.40E-04	7.47E-01	1.88E-01	2.48E-01
FA03	O108 H45	8.00	3.51	0.83	26.37	2.41	0.74	1.24E-06	9.98E-01	1.43E-01	1.54E-01

Transport parameters were determined by fitting (Fig. 7.4) a two-site non-equilibrium model to measured breakthrough data using HYDRUS 1D.

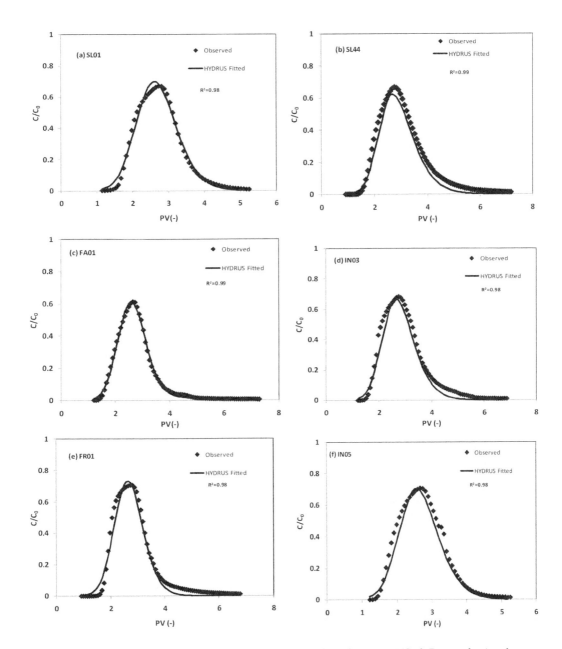

Fig. 7.4:   Examples of breakthrough curves fitted with HYDRUS 1-D to obtain the
              modeling parameters f, ω and k_s.

The resulting model parameters f, ω and k_s are given in Table 7.2, and with these, the
sticking efficiency, α was determined (Table 7.2). Even though the results indicated a
sticking efficiency variation of 5 log units between strain FR01 and strain IN05, the lowest

and highest attaching bacteria strains, respectively, we observed that 75 % of the strains had α values in the order of 10^{-3} and 10^{-4}.

In order to compare the importance of equilibrium sorption with kinetic attachment, we determined the total bacteria mass fraction involved in both types of interactions with the quartz grains. The total bacteria mass fraction involved in equilibrium sorption was determined by making use of equation (4). The total bacteria mass fraction involved in kinetic attachment was determined from the measured difference between total bacteria mass input and total bacteria mass output. The results suggested two groups of bacteria strains (Fig.7. 5):

1. A group of strains with approximately equivalent (low) mass fractions involved in both equilibrium sorption and kinetic attachment (group I), and
2. A group of strains with a high fraction of their cells involved in kinetic attachment, while the fraction of cells involved in equilibrium sorption was low and comparable to the group I strains (group II).

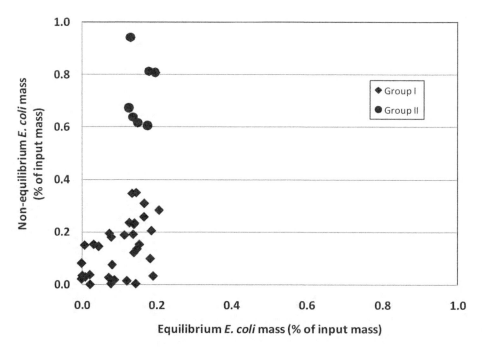

Fig. 7.5:   *Relation between fraction of cells with kinetic and equilibrium transport characteristics, indicating two groups:Group I-strains with equivalent (low) mass fractions involved in both equilibrium sorption and kinetic attachment and group II-strains with high fraction of cells involved in kinetic attachment*

Most of the strains (33 out of 40, or 82%) belonged to group I. Their breakthrough were somewhat retarded, while their maximum relative breakthrough concentrations were above 0.5; their sticking efficiency values were usually between 10^{-3} and 10^{-4}. The group II strains

(7 out of 40, or 18%) were all subject to considerable kinetic attachment, and their maximum relative breakthrough concentrations were therefore rather low: between 0.06 and 0.48.

7.4.4. Correlation between cell properties and model parameters

The relations between measured cell properties, parameters obtained from HYDRUS modeling and also maximum peak relative breakthrough concentrations were analysed using Spearman's Correlation Test (Table 7.3).

Table 7. 3:   Correlation matrix for $(C/C0)max$, measured cell properties, and model parameters (α, f, ω, ks)

	$C/C_0)_{max}$	α	f	ω	k_s	Cell aggregation	Motility	Zeta potential	Hydrophobicity
Cell aggregation	-0.180	-0.059	-0.052	-0.023	0.219				
Motility	-0.059	-0.024	0.018	-0.099	0.054	0.154			
Zeta potential	0.215	0.368	-0.179	0.394	-0.116	-0.068	-0.369		
Hydrophobicity	0.048	-0.284	0.269	-0.275	-0.044	-0.023	-0.092	-0.046	
Average Size	-0.071	0.117	-0.263	0.073	-0.009	-0.080	0.074	-0.257	-0.050

From Table 7.3, we concluded that there was no statistically significant correlation between neither of the measured cell characteristics nor between any of the measured cell characteristics and the parameters obtained from modeling. Four identified O-serogroups were grouped and one way ANOVA was applied to compare mean values of transport characteristics and measured cell properties between the groups. Results indicated that with the exception of f which showed a significant ($p = 0.021$) different mean value between serogroups O21 and O108 mean values of transport characteristics for the serogroups (O14, O21, O91 and O108) were not significantly different between the groups, with high p values ($p > 0.05$) (for α, $p = 0.254$; for ω, $p = 0.605$; for (C/C_0) max. $p = 0.657$; for k_s, $p = 0.833$; for k_a, $p = 0.276$). Average values of cell properties grouped according to the four O-serogroups did not show any significant difference among their mean values (for cell aggregation, $p = 0.531$; Motility, $p = 0.156$, zeta potential, $p = 0.534$; hydrophobicity, $p = 0.483$, size, $p = 0.818$).

7.5 Discussion

An important conclusion from our work was that almost all springs we sampled had high concentrations of thermotolerant coliforms, nitrate and chloride, whereby nitrate and chloride were correlated. This suggested that waste water, which is so abundantly present and disposed of freely in Kampala was the source of contamination of the springs. Our findings are not new: also Howard et al (2003) and Kulabako et al., (2007) identified multiple sources of anthropogenic contamination of the shallow aquifer from which most of the springs tap their water: solid waste dumps, pit latrines, unlined grey water channels or grey water polluted unlined storm water drainage channels are all prevalent in most of the Kampala area. Also, spring protection was not optimal: protective fences were broken and faulty allowing free-range cattle and other domestic animals to access the site and thereby increasing the vulnerability of the springs to fecal contamination. The lack of direct correlation between EC and thermotolerant coliforms suggested the possibility of multiple sources of fecal pollution leaching into the aquifers and polluting ground- and springwaters.

Based on the column experiments, an important second conclusion from our work was that maximum peak relative breakthrough and modeling results revealed an overall transport homogeneity among the *E. coli* strains: some 82 % of the strains had a maximum relative breakthrough concentration between 0.5 and 1, while some 75% of the 40 strains we tested had similar attachment efficiency values in the order of 10^{-3} and 10^{-4}. Such observed sticking efficiency homogeneity is in sharp contrast with our previous works (Foppen et al., 2010; Lutterodt et al., 2009a,b) and observations made by other workers (Schinner et al., 2010; Yang et al., 2008; Bolster et al., 2009, 2010), who observed significant variations in transport among the various *E. coli* strains used in their experiments. Though it is difficult to explain such observed homogeneity from *E. coli* surface characteristics (Table 7.3), the strains used in this work were isolated from flow lines at their termination points (springs), and such strains may therefore possess certain cell characteristics that might have influenced their selective transport in the subsurface giving rise to their similar transport characteristics in our columns.

In contrast to some workers (e.g. Bolster et al., 2006, 2010, Jacobs et al., 2007, Walker et al., 2005), who reported that hydrophobicity and zeta potential play an important role in bacteria transport, but consistent with our previous observations (Lutterodt et al., 2009a, Foppen et al., 2010) and observations made by Bolster et al. (2009, 2010), our results demonstrated that there was no statistically significant correlation between measured cell properties (zeta potential, motility, cell size, cell aggregation, and hydrophobicity) and transport parameters (f, ω and k_s and $(C/C_0)_{max}$). Bolster et al. (2010) explained that additional confounding factors may be present when correlating *E. coli* transport to surface properties. In this work, the lack of correlation between cell properties and bacteria attachment can be ascribed to the homogeneity in α and the observed large variation in cell properties. In contrast to Foppen et al. (2010), there was no correlation between α and $(C/C_0)_{max}$. We attributed this to the fact that all strains were not only subject to kinetic attachment, but also to equilibrium sorption. This type of sorption was absent in the *E. coli* strains obtained from the Rotterdam Zoo in the work of Foppen et al. (2010). In addition, Foppen and Schijven (2006) reported that, based on a number of studies, equilibrium sorption in *E. coli* transport was of little significance (Pang et al., 2003; Powelson and Mills, 2001; Sinton et al., 1997; Sinton et al., 2000; Alexander and Seiler, 1983; Havemeister and Riemer, 1985, both in: Matthess et al., 1985 and in Matthess et al., 1988; Champ and Schroeter, 1988; Merkli, 1975). Although the role equilibrium sorption plays in most of the strains we used in this work was indeed also rather limited, the role of kinetic attachment was at least equally limited. We do not have a good explanation for this equilibrium sorption phenomenon we observed. Possibly it was inherent to the approach we took in collecting *E. coli* strains: isolated from flow lines at their termination points (springs), and these strains may therefore have possessed certain typical transport characteristics that became apparent in our columns.

Another conclusion from our work was that 58% of the strains were of the O21:H7 serotype. The O antigen of *E. coli* consists of many repeats of an oligosaccharide unit and forms part of the lipopolysaccharide (LPS) molecule protruding from the outer membrane of an *E. coli* cell into its immediate environment. The O antigen determines the serogroup of an *E. coli* and the specific combination with the flagellar (H) antigen determines the serotype of an isolate (Stenutz et al., 2006). LPS is anchored in the outer membrane of *E. coli* and occupies 75% of the surface of the gram-negative bacterium, and *E. coli* is known to have some 10^6 LPS molecules per cell (Caroff and Karibian, 2006). The interaction of *E. coli* with its immediate surrounding may to a cetain extent therefore be controlled by these molecules. For instance, Foppen et al. (2010) found an association between *E. coli* serotype, on one hand, and attachment efficiency on the other hand. Though the transport mechanisms in the Kampala aquifers of the strains we used in this work are unknown, the specific serotype O21:H7 may possibly possess certain characteristics that allow its preferential transport through the aquifers in the Kampala area, and probably some 60% of the strains possessed this particular serotype.

According to Garabal et al. (1996), *E. coli* strains belonging to a specific serotype or serogroup do not themselves confer virulence, rather, it has often been demonstrated that serotyping can be used as an indicator for pathogenicity, since there is a high positive correlation between certain serotypes and (entero)pathogenicity. In this work, five serotypes were identified: O21:H7, O14:H15, O14:H4, O91:H28, O108:H45, and a group with O-undefinable serotypes. It is interesting to note that all definable O-serogroups were associated with (diarrheal) diseases: Jenkins et al. (2006) and Kang et al. (2001) reported isolation of *E. coli* O21 from patients with diarrhea, while Knobl et al. (2001) also identified a pathogenic *E. coli* isolate of serogroup O21. Furthermore, *E. coli* belonging to O14 and O91 serogroups are known to be associated with urinary tract infections and diarrhea with life threatening complications (Stenutz et al., 2006, Bettelheim, 1978). Finally, *E. coli* O108 has been isolated from faeces of adults with urinary tract infections (Bettelheim, 1978).

7.6 Conclusions

Based on the work from the Kampala springs, the following conclusions can be drawn:
- Almost all springs we sampled had high concentrations of thermotolerant coliforms, nitrate and chloride, whereby nitrate and chloride were correlated. This suggested that waste water, which is so abundantly present and disposed of in Kampala, was the source of contamination of the springs.
- Based on column experiments, we concluded that the transport of the *E. coli* strains was remarkable similar: some 82 % of the strains had a maximum relative breakthrough concentration between 0.5 and 1, while some 75% of the 40 strains we tested had similar attachment efficiency values in the order 10^{-3} and 10^{-4}. We attributed this to the way in which the strains were harvested: from springs, and therefore at the termination points of flowlines. Such strains may therefore possess certain cell characteristics that might have influenced their selective transport in the subsurface giving rise to their similar transport characteristics in our columns.
- There was, however, no statistically significant correlation between measured cell properties (zeta potential, motility, cell size, cell aggregation, and hydrophobicity) and transport parameters (f , ω and k_s and $(C/C_0)_{max}$).
- The transport of all strains was not only affected by kinetic attachment, but also by equilibrium sorption. For this, we do not have a good explanation for this equilibrium sorption phenomenon we observed.
- Of the strains we tested 58% were of the O21:H7 serotype. This suggests that the specific serotype O21:H7 may possess certain characteristics that allow its preferential transport through the aquifers in the Kampala area.

- In this work, five serogroups were identified: O21:H7, O14, O91, O108, and a group with O-undefinable serotypes. In literatre, all these definable O-serotypes and serogroups have been associated with (diarrheal) diseases.

PART IV

SUMMARY, CONCLUSIONS AND
RECOMMENDATIONS

Chapter 8 Summary, conclusions and recommendations

8.1 Point of departure and scope

Groundwater, one of the most important sources of water for drinking water supplies in many regions of the world is under threat by microbial pathogenic contamination due to anthropogenic activities. An effective way of protecting the resource from contamination, especially, by pathogenic microorganisms leaking into an aquifer, is by delineating protection areas around a drinking source. Surface water can also be effectively treated by passage through sand to remove pathogens provided travel distances and times are adequate (e.g. Tufenkji et al., 2003, Schijven, 2001). These strategies rely upon effective natural attenuation of microorganisms by soils over set back distances (e.g. Taylor et al., 2004), and it is dependent on the interaction between cells and aquifer media resulting in cell retention. Even though natural processes may assist in the reduction of pollution and it is widely utilized in soil aquifer treatment sites, most biological contaminants can travel long distances through soils and aquifers until they are discharged into streams or wells (Corapcioglu and Haridas, 1985).

To predict the presence of pathogens in water, a separate group of microorganisms, generally known as faecal indicator organism are used; one of the most important indicators used worldwide is *Escherichia coli*. Due to the importance of *E. coli* as faecal indicator bacteria, considerable attention has been given to understanding their transport and fate in saturated porous media (e.g. Foppen et al, 2007a.b, Schinner et al., 2010, Bolster et al., 2010). Even though column experiments have the disadvantage of the inability to simultaneously study the various factors that influence *E. coli* transport in the subsurface, a unique advantage associated with using columns is the provision of a great degree of control (Shani et al., 2008, Harvey and Harms, 2002), and this allows the possibility of isolating and studying specific factors affecting colloid transport. Results from column experiments are traditionally modeled using the CFT (Yao et al., 1971), which provides an important framework for predicting transport of *E. coli* in saturated porous media (Foppen, 2007). A unique characteristic of the theory is the application of the sticking efficiency to predict colloid transport distances in aquifers, the sticking efficiency is defined in Chapter 1 as the ratio of the rate of particles striking and sticking to a collector to the rate of particles striking a collector, and is mainly determined by electro-chemical forces between the colloid and the surface of the collector. According to the theory, the sticking efficiency is constant in time and distance (Yao et al., 1971; Tufenkji and Elimelech, 2004a).

Contrary to the CFT, research results have indicated that the sticking efficiency of a bio-colloid population varies due to variable surface properties of individual members of the population, resulting in differences in affinity for collector surfaces (Albinger et al., 1994; Baygents et al., 1998.; Simoni et al., 1998; Li et al., 2004; Tufenkji and Elimelech, 2005a; Tong and Johnson, 2007; Foppen et al., 2007a) and contribute to distributions in bacteria attachment efficiency.

This research involved the study of the transport of various *E. coli* strains isolated from different sources of the environment (feces and different parts of zoo animals, soils of a pasture used for animal grazing, and springs). In addition, prior to experiments, strains were grown in cow manure at 21°C (Chapters 2 and 4) to mimic environmental conditions (Yang et al., 2006) or in nutrient broth at 37°C (Chapters 3, and 5-7) comparable to intestinal conditions, the optimal growth temperature for *E. coli*. Experiments were conducted under laboratory controlled conditions with a constant range of quartz grain sizes saturated with low and high ionic strength solutions in columns of lengths up to 25 m. In addition, a constant fluid flow velocity (Darcy velocity) was maintained in all experiments. Short (7 cm) and long (1.5 - 25 m) columns were used to investigate inter-strain attachment variations among the strains, and multiple sampling distances along the lengths of the long columns were applied to study intra-strain attachment differences, distributions in attachment efficiency within *E. coli* strains and to develop a methodology to measure the minimum sticking efficiency, in addition, the long column experiments were used to measure low values of the distance dependent sticking efficiency. Cell properties, phenotypic characteristics (motility, average cell size, cell aggregation, hydrophobicity and zeta potential) and genes encoding structures at the outer membrane of *E. coli* cells were measured prior to experiments to investigate their effects on transport/attachment.

8.2 Variability in *E. coli* transport, low values of sticking efficiencies and the minimum sticking efficiency

Transport experiments in long columns with multiple sampling ports at increasing transport distances helped in studying intra-strain attachment variations, distributions in sticking efficiency and measuring low values of the distance dependent sticking efficiency of fractions of cells within *E. coli* strains. In addition, two computational methods that make use of relative bacteria mass breakthrough to quantify cell attachment (Abudalo et al., 2005, Kretzschmar et al., 1997) were applied. First, the entire transport distance from top (influent) of the column to a sampling port was considered, and this allowed for the comparison of transport of different *E. coli* strains at equal and increasing distances (Chapters 2 and 4 to 6). The second method involved the computation of sticking efficiency of cells retained in a column segment (Martin et al., 1996), in between two sampling distances. In this way, the sticking efficiency of fraction of total mass input retained in a segment could be determined and distribution functions that best described the relation between the two parameters were assessed (Chapters 4 to 6). Short column experiments were conducted for strains isolated from zoo animals, soils of a pasture used for animal grazing (Chapter 3) and for strains isolated from springs in the Kampala area in Uganda (Chapter 7).

The relation between fraction of cells retained in a column segment and corresponding sticking efficiency was best described by a power-law (Chapters 4-6), though exponential distribution was equally good (Chapter 4). In chapters 4 and 5, the minimum sticking efficiency was introduced and defined as the sticking efficiency belonging to a bacteria fraction of 0.001% of initial bacteria mass flowing into a column, after removal of 99.999% (5 log reduction) of the original bacteria mass has taken place. This minimum sticking efficiency was extrapolated from the power-law relation between segment sticking efficiencies and mass fractions retained in segments. However, within this minor fraction of bacteria cells, the sticking efficiencies were not a constant but distributed, and within this 0.001% sub-fraction, the minimum sticking efficiency is the *highest* possible sticking efficiency.

From results obtained in all experiments, it was concluded that intra-strain and inter-strain heterogeneities existed within and among the *E. coli* strains studied. Inter-strain attachment variations were observed in all transport experiments conducted: In chapter 3, *E. coli* strains isolated from soils and from feces and different parts of zoo animals showed a two log unit variation in maximum breakthrough, whereas experiments with *E. coli* strains isolated after transport through springs (Chapter 7) resulted in overall homogeneity. The majority of the strains exhibited remarkably similar transport characteristics, with some 82 % having a maximum relative breakthrough concentration between 0.5 and 1, while some 75% of the 40 *E. coli* strains had similar attachment efficiency values in the order 10^{-3} and 10^{-4}. This is even though the attachment variation between two strains with the least (10^{-7}) and highest (10^{-2}) sticking efficiencies resulted in a 5 log unit variation. The observed homogeneity is attributable to the way in which the strains were harvested: from springs, and therefore at the termination points of flow lines. Such strains may therefore possess certain cell characteristics that might have influenced their selective transport in the subsurface giving rise to their similar transport characteristics in our columns.

Intra-strain attachment variations were observed for all *E. coli* strains used, regardless of the source of isolation and growth medium. A general trend of reduction in sticking efficiencies with increasing transport distances was observed. A summary of the long column results is presented in Table 8.1.

Table 8.1: *Lowest measured values of sticking efficiencies (α_L, α_i) values and extrapolated minimum sticking efficiency (α_{min}) for long column (L ≥1.47) transport experiments in artificial groundwater*

Strain	Source of isolation	Growth medium	Column length (m)	Lowest α_L (-)	Lowest α_i (-)	α_{min} (-)	Fraction exiting the column (-)
UCFL-71	Soils	manure extract	4.83	2.6×10^{-1}	8.5×10^{-2}	4.7×10^{-2}	4.5×10^{-6}
UCFL-94	Soils	manure extract	4.83	2.5×10^{-1}	6.0×10^{-3}	6.1×10^{-6}	2.1×10^{-1}
		nutrient broth	25.65	1.2×10^{-3}	8.2×10^{-6}	1.7×10^{-7}	1.9×10^{-1}
UCFL-131	Soils	manure extract	4.83	3.5×10^{-1}	2.8×10^{-1}	9.4×10^{-3}	$<1 \times 10^{-6}$
		nutrient broth	25.65	2.3×10^{-4}	1.6×10^{-5}	6.8×10^{-8}	8.7×10^{-1}
UCFL-167	Soils	manure extract	4.83	3.5×10^{-1}	1.4×10^{-1}	1.7×10^{-1}	$<1 \times 10^{-6}$
UCFL-263	Soils	manure extract	4.83	3.6×10^{-1}	2.0×10^{-1}	3.4×10^{-2}	$<1 \times 10^{-6}$
UCFL-348	Soils	manure extract	4.83	>1	>1	≥ 1	n.d.
SL03	Spring	nutrient broth	1.47	2.8×10^{-2}	1.2×10^{-2}	1.3×10^{-5}	3.4×10^{-1}
FR08	Spring	nutrient broth	1.47	2.5×10^{-2}	1.3×10^{-2}	1.7×10^{-4}	4.2×10^{-1}
FR02	Spring	nutrient broth	1.47	1.1×10^{-2}	2.0×10^{-3}	1.8×10^{-4}	2.0×10^{-2}
FA03	Spring	nutrient broth	1.47	5.5×10^{-2}	1.0×10^{-2}	8.1×10^{-5}	1.3×10^{-1}
SL20	Spring	nutrient broth	1.47	1.5×10^{-2}	3.0×10^{-3}	4.7×10^{-5}	6.6×10^{-1}
FR05	spring	nutrient broth	1.47	9.3×10^{-2}	7.0×10^{-2}	~0.1	3.0×10^{-2}

Both segment sticking efficiency (α_i) and sticking efficiency measured over total transport distances (α_L) reduced with increasing column lengths and indicated intra-strain attachment variations for all strains used for our long column transport experiments. In addition, we concluded that environmentally relevant low values of sticking efficiencies in the order 10^{-3} to 10^{-6} were measurable in the laboratory, and the results demonstrated the importance of the use of long columns. The measured low values of sticking efficiency show that, for bacterial populations leaked into groundwater environments, sub-populations may posess non-attaching characteristics and therefore increases their chances of being transported over considerable distances.

Differences in cell attachment to quartz grains resulted in power-law distributions describing the relation between *E. coli* sub-populations and corresponding sticking efficiencies. From power-law equations, the minimum sticking efficiency defined above and in Chapters 3 and 4 were obtained by extrapolations. The minimum sticking efficiency gives the α_i-values of lower end fractions of cells in the power-law distribution with non-attaching characteristics. Such cells may be transported over distances much longer than predicted, if results from short column experiments are extrapolated to the field scale. The low values of extrapolated minimum sticking efficiencies compared with measured values of sticking efficiencies over the total transport distances and column segments (Table 8.1) makes the minimum sticking efficiency a valuable tool in delineating well-head protection areas in real-world scenarios.

8.3 Effect of cell characteristics on the transport/attachment of *E. coli* cells

Results from the experiments revealed that none of the studied phenotypic characteristics (motility, hydrophobicity, cell aggregation, cell size, cell surface charge i.e. zeta potential or outer surface potential) promoted *E. coli* retention on the quartz grains in solutions of high ionic strength. Though cell motility and Antigen-43 (Ag43) expression significantly influenced *E. coli* attachment to quartz grains over relatively short transport distances in low ionic strength solutions (chapter 2), under a more relevant groundwater chemistry (solution of magnesium sulphate and calcium chloride) in chapters 3 and 7, results indicated a lack of correlation between bacteria attachment efficiency and the two *E. coli* characteristics (motility and Ag43-expression).

Though the transport and *E. coli* characterization experiments in chapters 3 and 7, respectively, revealed a non-significant correlation and a non-correlation of *E. coli* serogroup with transport characteristics, 58% of all spring *E. coli* isolates that were serotyped belonged to the same serogroup O21 (serotype O21:H7). It is therefore speculated that the specific serotype O21:H7 may possess certain characteristics that promote their selective transport through the aquifers in the Kampala area in Uganda. The homogeneity in transport characteristics of the spring isolates together with high percentage of strains belonging to the same serotype that were transported through the springs of Kampala and the two log unit variation among *E. coli* isolates in chapter 3 (from soils and zoo animals) indicated that, for various bacteria strains used in laboratory controlled experiments, the source of isolation may have a significant influence on the transport characteristcs especially when travel distances are short (<10 cm).

8.4 Recommendations for future research

Although this research unraveled the distribution in sticking efficiency at large inter-transport distances, and therefore added to the evidence that the classical colloid filtration theory is over-simplified when applied to predict bio-colloid transport distances in saturated porous media, the effect of cell properties on the transport and attachment of bacteria strains to quartz sand (a predominant aquifer media) still remains a mystery. This is in spite of the fact that the variability in transport and attachment characteristics within and among different *E. coli* strains have been attributed to different cell surface properties or a complex combination of cell surface organelles. It is therefore important for future research to focus on cell surface structures known to be involved in initial attachment to host tissues and/or abiotic surfaces. This will involve the identification of the relative contribution of various cell surface structures to bacterial attachment to aquifer media. To achieve this, a number of genes known to be involved in initial attachment need to be targeted; these genes include fimH, afa, ompC, slp, and surA, Also, the so-called Keio collection of single gene deletion mutants of all non-essential genes of *E. coli* K-12 is recommended. The Keio collection uses *E. coli* K-12, and therefore, the arrangement and expression of surface structures and their combined effect on initial attachment is characteristic for *E. coli* K-12, but not necessarily characteristic for initial attachment of (wild) *E. coli* strains found in the environment. It will therefore be important to identify the presence of some (2 or 3) of the most important surface structures on wild *E. coli* strains for environmentally relevant conditions. The expression of surface structures is usually governed by complex pathways, involving a wide variety of genes. For instance, for the assembly of the motor of flagella of *E. coli* , the controlled expression of about 50 genes is required. The control is governed by various switch proteins and mechanisms, which, on their turn, are influenced by the environment the cell is in.

To further understand the contribution of various cell surface structures on their transport and retention in aquifers, future transport experiments must focus on field bacterial transport experiments. This will offer a better understanding of the various environmental factors that may affect bacterial transport in real-world environments.These factors may include hydrochemical and geochemical controls excerted on transport, like DOC, presence of iron-oxides, sedimentary organic carbon, and calcium carbonates. In addition, the role of dual porosity or perhaps even fissure flow types of environments may be more adequately be assessed under field conditions than under laboratory conditions.

List of Symbols

ϖ	Absolute fluid viscosity ($ML^{-1}T^{-1}$)
ε_0	Dielectric permittivity in a vacuum ($CV^{-1}m^{-1}$)
g	Acceleration due to gravity (Lt^{-2})
k_a	Attachment rate coefficient (t^{-1})
k	Boltzmann's constant (JK^{-1})
C	Cell suspension or concentration (# cells L^{-3})
ε_r	Dielectric constant (-)
x	Distance (L)
μ	Electrophoretic mobility ($L^2V^{-1}t^{-1}$)
k_v	Equilibrium sorption coefficient (L^3M^{-1})
ω	First order rate constant (t^{-1})
U	Fluid approach velocity (Lt^{-1})
f	Fraction of exchange sites assumed to be in equilibrium with solution phase (-)
S_e	Mass adsorbed at equilibrium sites (# cells)
S_k	Mass adsorbed at kinetically controlled sites (# cells)
d_c	Median of the grain size weight distribution (L)
ψ_0	Outer Surface Potential (V)
v	Pore water flow velocity (Lt^{-1})
α_i	Segment sticking efficiency (-)
η_0	Single collector contact efficiency (-)
η	Single collector removal efficiency (-)
α	Sticking or attachment efficiency (-)
η	The fluid dynamic viscosity (Pa s)
t	Time (s).
θ	Total porosity of the sand (-)
S	Total retained bacteria concentration (#cellsM^{-1})
x	Transport distance (L)
L	Travel distance or length of column (L)
q	Volumetric flow rate (L^3t^{-1})
N_R	Aspect ratio or interception number (-)
N_A	Attraction number (-)
$\bar{\alpha}_{strain}$	Average sticking efficiency over total transport distances (-)
ρ_{bulk}	Bulk density (ML^{-3})
ρ_{fix}	Fixed charge density (mol)
ρ_f	Fluid density (ML^{-3})
N_G	Gravitation number (-)
H	Hamaker constant between collector and colloid (J)
A_S	Happel's cell model constant (-)
d_p	Mean Collector grain diameter (L)

d_p	Mean Particle grain diameter (L)
α_{min}	Minimum sticking efficiency (-)
ρ_P	Particle density (ML^{-3})
N_{Pe}	Peclet number (-)
α_L	Sticking efficiency over total transport distance (-)
$\frac{1}{\kappa}$	The double layer thickness (L)
$\frac{1}{\lambda}$	The electrophoretic softness (L)
M_{eff}	Total number of cells in the effluent (# cells)
M_{inf}	Total number of cells in the influent (# cells)
N_{vdW}	van der Waals attraction number (-)
η_D	Theoretical values for the single collector contact efficiency when the sole transport mechanism is diffusion (-)
η_I	Theoretical values for the single collector contact efficiency when the sole transport mechanism is by interception (-)
η_G	Theoretical values for the single collector contact efficiency when the sole transport mechanism is gravitational settling (-)
D, D_B	Hydrodynamic dispersion coefficient (L^2t^{-1})
T	Temperature (K)

Abbreviations

AGW	Artificial Groundwater
ANOVA	Analysis of Variance
CFT	Colloid Filtration Theory
DI	De-mineralized or De-ionized Water
DNA	Deoxyribonucleic acid
EC	Electrical Conductivity
EHEC	Entero-hemorrhagic *Escherichia coli*
LPS	Lipopolysaccharides
MATH	Microbial Adhesion to Hydrocarbon
MDG	Millenium Development Goal
PCR	Polymerase Chain Reaction
PV	Pore Volume
PVC	Polyvinyl chloride
SCCE	Single collector contact efficiency
SCRE	Single collector removal efficiency
UCFL	University of Connecticut Feedlot
WHO	World Health Organisation

References

Abudalo, A.R., Bogatsu, Y.G., Ryan, J.N., Harvey, R.W. Metgel. D.W. and Elimelech, M. (2005) Effect of ferric oxyhydroxide grain coatings on the transport of Bacteriophage PRD1 and *Cryptosporidium parvum Oocysts* in saturated porous media. Environ Sci Technol., 39: 6412-6419.

Albinger, O., Biesemeyer, B.K., Arnold, R.G., and Logan, B.E. (1994) Effect of bacterial heterogeneity on adhesion to uniform collectors by monoclonal populations. FEMS Microbiol. Letters, 124: 321-326.

Alexander, I., and Seiler, K.P. (1983) Lebensdauer und Transport von Bakterien in typischen Grundwasserleitern. Munchener Schotterebene. DVGW-Schr., 35: 113-125 (ZfGW, Frankfurt/Main).

Baxter, K.M. and L. Clark, (1984) Effluent recharge. The effects of effluent recharge on groundwater quality. Technical Report 199. Water Research Centre. United Kingdom.

Baygents, J.C., Glynn, J.R., Albinger, O., Biesemeyer, B.K., Ogden, K.L., and Arnold, R.G. (1998) Variation of surface charge density in monoclonal populations: Implications for Transport through Porous Media. Environ. Sci. Technol., 32: 1596-1603.

Becker, M.W., Collins, S.A., Metge, D.W., Harvey, R.W., and Shapiro, A.M. (2004) Effect of cell physicochemical characteristics and motility on bacterial transport in groundwater. J. Contam. Hydrol., 69: 195-213.

Beloin, C., J. Valle, P. Latour-Lambert, P. Faure, M. Kzreminski, D. Balestrino, J.A. Haagensen, S. Molin, G. Prensier, B. Arbeille, J.M. Ghigo, (2004) Global impact of mature biofilm lifestyle on *Escherichia coli* K-12 gene expression, Mol. Microbiol., 51 : 659–674.

Benekos, I. D., Cirpka, O. A., and Kitanidis, P. K. (2006) Experimental Determination of transverse dispersivity in a helix and a cochlea, Water Resources Research., 42, W07406, doi: 10.1029/2005WR004712.

Betelheim, K.A. (1978) The sources of 'OH' serotypes of *Escherichia coli*. J. Hyg., Camb. 80: 83-113.

Bhattacharjee, S., Ryan, J.N., and Elimelech, M. (2002).Virus transport in physically and geochemically heterogeneous subsurface porous media. J. Contam. Hydrol., 57: 161-187.

Bolster ,C.H., Cook, K. L, Marcus, I. M, Haznedaroglu, B. Z., and Walker, S. L. (2010) Correlating Transport Behavior with Cell Properties for Eight Porcine *Escherichia coli* Isolates. Environ. Sci. Technol., 44: 5008–5014.

Bolster, C. H., Haznedaroglu, B. and Walker, S. L. (2009) Diversity in cell properties and transport behaviour among 12 different environmental *Escherichia coli* isolates. Journal of Environmental Quality, 38: 465-472.

Bolster, C.H., Mills, A.L., Hornberger, G. M. and Herman, J. (1999) Spatial distribution of deposited bacteria following miscible displacement experiments in intact sediment core. Water Resources Research ,35: 1797-1807.

Bolster, C.H., Mills, A.L., Hornberger, G. M. and Herman, J. (2000) Effect of intra-population variability on the long distance transport of bacteria. Ground Water , 38: 370-375.

Bolster, C.H., Mills, A.L., Hornberger, G.M., and Herman, J.S. (2001) Effect of surface coatings, grain size, and ionic strength on maximum attainable coverage of bacteria on sand surfaces. J. Contam. Hydrol., 59: 287-305.

Bolster, C.H., Walker, S.L., and Cook, K.L. (2006) Comparison of *Escherichia coli* and *Campylobacter jejuni* Transport in saturated porous media. J. Environ. Qual., 35: 1018-1025.

Boyce, T.G., D.L. Swerdlow, and Griffin, P.M. (1995) *Escherichia coli* O157:H7 and the hemolytic-uremic syndrome. N Engl J Med 333: 364–368.

Bradford, S., and Bettahar, M. (2005) Straining attachment and detachment of *Cryptosporidium Oocysts* in saturated porous media. J. Environ. Qual., 34:469-478.

Bradford, S., Torkzaban. S, and Walker, S.L. (2007). Coupling physical and chemical mechanisms of colloid straining in saturated porous media. Water Res., 41:3012-3024.

Bradford, S.A., J. Simunek, and S.L. Walker, (2006) Transport and straining of *E. coli* O157:H7 in saturated porous media. Water Resour. Res., 42, W12S12, doi: 10.1029/2005WR004805.

Bradford, S.A., Simunek, J. Bettahar, M., van Genuchten, M.T., and Yates, S.R. (2003) Modeling colloid attachment, straining and exclusion in saturated porous media. Environ Sci. Technol., 37: 2242-2250.

Bradford, S.A., Yates, S.R., Bettahar, M., and Simunek, J. (2002) Physical factors affecting the transport and fate of colloids in saturated porous media. Water Res. Res., 38:1327-1340.

Brown, D. G., and Abramson.A (2006) Collision efficiency distribution of a bacterial suspension flowing through porous media and implications for field transport.Water Research, 40:1591-1598.

Byamukama, D., Kansiime, F., Mach, R.L., and Farnleitner, A.H.(2000) Determination of *Escherichia coli* contamination with chromucult coliform agar showed a high level of discrimination efficiency for differing fecal pollution levels in tropical of Kampala, Uganda. Applied and Environmental Microbiology, 66: 864-868.

Cameron, D.R., and Klute, A. (1977) Convective-dispersive solute transport with a combined equilibrium and kinetic adsorption model. Water Res. Res., 13:183-188.

Canter, L.W. and Knox, R.C. (1985) Septic tank system effects on groundwater quality. Lewis Publishers, Inc., Chelsea, Michigan USA. ISBN 0-87371-012-6.

Caroff, M., and Karibian, D (2003) Structure of bacterial lipopolysaccharides. Carbohydrate Research, 338: 2431-2447.

Champ, D.R. and Schroeter, J. (1988) Bacteria transport in fractured rock- A field scale tracer test at the Chalk River Nuclear Laboratories. Water Sci. Technol., 20: 81-87.

Chapman, T.A., Wu, X-Y., Barchia, I., Bettelheim, K.A., Driesen, S. Trott, D., Wilson, M and Chin, J.J.-C (2006) Comparison of Virulence Gene Profiles of *Escherichia coli* Strains Isolated from Healthy and Diarrheic Swine. Applied Environmental Microbiology, 72, 4782-4795.

Cirpka, O. A. and Kitanidis, P. K (2001) Theoretical basis for the measurement of local transverse dispersion in isotropic porous media. Water Resources Research, 37: 243-252.

Close, E.M., Pang, L., Flintoft, M.J. and Sinton, L.W.(2006) Distance and flow effects on microsphere transport in a large gravel column. Journal of Environmental Quality, 35:1204-1212.

Corapcioglu, M.Y and Haridas, A (1984) Transport and fate of microorganisms in porous media: a theoretical investigation. J. Hydrol., 72: 149-169.

Corapcioglu, M.Y and Haridas, A (1985) Microbial transport in soils and groundwater: a numerical model. Adv. Water Resour., 8: 188-200

Dague, E., Duval, J., Jorand, F., Thomas, F., and Gaboriaud, F. (2006) probing surface structures of *Shewanella spp.* by microelectrophoresis. Biophysical J, 90: 2612-2621.

Danese, P.N., Pratt, L.A., Dove, S.L., and Kolter, R. (2000) The outer membrane protein antigen 43 mediates cell-to-cell interactions within *Escherichia coli* biofilms. Mol. Microbiology, 37: 424-432.

Das Gracas de Luna1, M., Scott-Tucker, A., Desvaux, M., Ferguson,, P. , Morin, N. P., Dudley, E. G., Turner, S., Nataro, J. P., Owen, P and Henderson, I. R. (2008) The *Escherichia coli* biofilm formation -promoting protein Antigen43 does not contribute to intestinal colonization. FEMS Microbiol Lett, 284:237-246.

de Kerchove, A. J., and Elimelech, M. (2008) Calcium and magnesium cations enhance the adhesion of motile and nonmotile *Pseudomonas aeruginosa* on alginate films. Langmuir, 24: 3392-3399.

Dong, H., Rothmel, R., Onstott, T.C., Fuller, M.E., DeFlaun, M.F., Stretger, S.H., Dunlap, R., and Fletcher, M. (2002) Simultaneous transport of two bacterial strains in intact cores from Oyster, Virginia: Biological effects and numerical modeling. Appl. Environ. Microbiol., 68:2120-2132 .

Dong, H., Scheibe, T.D., Johnson, W. P., Monkman, C.M. and Fuller, M. E. (2006) Change of collision efficiency with distance in bacterial transport experiments. Ground Water ,44: 415-429.

Elimelech, M., Nagai, M., Ko, C-H., and Ryan, J.(2000) Relative insignificance of mineral outer surface potential to colloid transport in geochemically heterogeneous porous media. Environ. Sci. Technol., 34:2143-2148.

Ewing, W.H. 1986. Edwards and Ewing's Identification of Enterobacteriaceae. 4th Edition. Elsevier Publishing Company, New York (p130).

Ezawa, A., Gocho, G., Saitoh, M., Tamura, T., Kawata, K., Takahashi, T and Kikuchi, N. (2004) A three-year study of enterohemorrhagic *Escherichia coli* O157 on a farm in Japan. J Vet Med. Sci., 66: 779–784.

Foppen, J. W., Lutterodt, G., Roling, W.F.M and Uhlenbrook, S. (2010) Towards understanding inter-strain attachment variations of *Escherichia coli* during transport in saturated porous quartz sand. Water Research, 44: 202-1212.

Foppen, J.W. (2007) Transport of *Escherichia coli* in saturated porous media. Ph.D. Thesis. ISBN 13-978-0415-44477-4. Taylor and Francis group plc, London ,UK .

Foppen, J.W.A., and Schijven, J.F. (2005) Transport of *E. coli* in columns of geochemically heterogeneous sediment. Water Research, 39: 3082-3088.

Foppen, J.W.A., Schijven, J.F., (2006). Evaluation of data from the literature on the transport and survival of *Escherichia coli* and thermotolerant coliforms in aquifers under saturated conditions. Water Res., 40: 401–426.

Foppen, J.W.A., van Herwerden, M., and Schijven, J.F. (2007a) Transport of *Escherichia coli* in saturated porous media: dual mode deposition and intra-population heterogeneity. Water Research ,41: 1743-1753.

Foppen, J.W.A., van Herwerden, M., Schijven, J.F., (2007b) Measuring and modeling of straining of *Escherichia coli* in saturated porous media. J. Contam. Hydrol., 93: 236–254.

Frank, B.P., and Belfort, G. (2005) Polysaccharides and sticky membrane surfaces: critical ionic effects. Journal of Membrane Science, 212: 205-212.

Gaboriaud, F., Dague, E., Sidney, B., Jorand, F., Duval, J and Thomas, F. (2006) Multiscale dynamics of the cell envelope of *Shewanella putrefaciens* as a response to pH change. Colloid and Surfaces B: Biointerfaces, 52:108-116.

Gagliardi, J.V. and Karns, J.S. (2002) Persistence of *Escherichia coli* O157:H7 in soil and on plant roots. Environmental Microbiology, 4: 89-96.

Gannon, J., Tan, Y., Baveye, P., and Alexander, M. (1991) Effect of sodium chloride on transport of bacteria in a saturated aquifer material. Appl. Environ. Microbiol., 57: 2497-2501.

Gerba, C.P. and Smith, Jnr J.E. (2005) Sources of pathogenic microorganisms and their fate during land application of wastes. Journal of Environmental Quality, 34: 42-48.

Gilbert, P., Evan, D.J., Evans, E., Duguid, I.G., and Brown, M.R. (1991) Surface characteristics and adhesion of *Escherichia coli* and *Stypahlococcus epidermis*. Journal of Appl. Bacteriology, 71:72-77.

Goss, M.J., Barr, D.A.J. and Rudolph, D.L. (1998) Contamination in Ontario farmstead domestic wells and its association with agriculture: Result from drinking water wells. J. Contam Hydrol., 32: 267-293.

Grimm, L.M., M. Goldoft, J. Kobayashi, J.H. Lewis, D. Alfi, A.M. Perdichizzi, P.I. Tarr, J.E. Ongerth, S.L. Moseley, and Samadpour, M. (1995) Molecular epidemiology of a fast-food restaurant-associated outbreak of *Escherichia coli* O157:H7 in Washington State. J Clin Microbiol 33: 2155–2158.

Guinée, P.A.M., Agterberg, C.M., Jansen, W.J. (1972) *Escherichia coli* O Antigen Typing by Means of a Mechanized Micro technique American Society for Microbiology 24: 127-31.

Harvey, R.W. and Harms, H. (2002) Transport of microorganisms in the terrestrial subsurface: in situ and laboratory methods. In: C.J. Hurst, G.R. Knudsen, M.J. McInerney, L.D. Stetzenback and R.L. Crawford, Editors, *Manual of Environmental Microbiology* (second ed), ASM Press, Washington, DC (2002), pp. 753–776.

Hasman, H., Schembri, M.A., and Klemm, P. (2000) Antigen 43 and type 1 fimbriae determine colony morphology of *Escherichia coli* K-12. J. Bacteriol., 182: 1089-1095.

Havemeister,G., and Riemer,R.(1985) Laborversuche zum Transport-verhalten von Bakterien. Umweltbundesambt Materialien, 2/85: 27-32

Haznedaroglu, B. Z., Bolster C.H. and Walker, S.L. (2008) The role of starvation on *Escherichia coli* adhesion and transport in saturated porous media. Water Research, 42: 1547-1554.

Henderson, I.A., Navarro-Garcia, F. and Nataro, J. P. (1998) The great escape: structure and function of the autotransporter proteins. Trends Microbiol. Sep; 6(9), p. 370-378.

Henderson, I.R., Meehan, M., and Owen, P. (1997) A phase variable bipartite outer membrane protein, determines colony morphology and autoaggregation in *Escherichia coli* K-12 FEMS Microbiol. Letters, 149: 115-120.

Howard, G., Pedley, S., Barrett., M., Nalubega, M., Johal, K. (2003) Risk factors contributing to microbiological contamination of shallow groundwater in Kampala, Uganda Water research, 37: 3421-3429.

Hrudey, S.E., Payment, P. , Huck, P.M., P.M. Gillham, R.W. , and Hrudey, E.J. (2003) A fatal waterborne disease epidemic in Walkerton, Ontario: comparison with other waterborne outbreaks in the developed world. Water Science and Technology, 47: 7-14.

Jacobs, A., Lagolie, F., Herry, J.M., and Debroux, M. (2007) Kinetic adhesion of bacterial cells to sand: Cell surface properties and adhesion rate. Colloid Surf. B Interfaces, 59:35-45.

Jenkins C, Chart H, Willshaw GA, Cheasty T, and Smith HR (2006) Genotyping of enteroaggregative *Escherichia coli* and identification of target genes for the detection of both typical and atypical strains. Diagnostic Microbiology and Infectious Disease, 55 13-19.

Jiang, G., Noonan, M.J., Buchan, G.D., Smith, N., (2007) Transport of *Escherichia coli* through variably saturated sand columns and modeling approaches. J. Contam. Hydrol., 93: 2–20.

Johnson, P.R., Sun, N., and Elimelech, M. (1996) Colloid transport in geochemically heterogeneous porous media; modeling and measurements. Environ. Sci. Technol., 30: 3284-3293.

Kaper, J.B., (1998). Enterohemorrhagic *Escherichia coli*. Curr Opin Microbiol 1: 103–108.

Kaper, J.B., J.P. Nataro, and Mobley, H.L. T. (2004) Pathogenic *Escherichia coli*. Nature Reviews Microbiology, 2: 123-140.

Klemm, P., Hjerrild, M., Gjermansen, M., and Schembri, M.A. (2004) Structure function analysis of the self recognizing antigen-43 autotransporter protein from *Escherichia coli*. Mol. Microbiology, 51: 283-296.

Knobl, T., Baccaro, M. R., Moreno, A.M., Gomes, T.A., Vieira, M.A., Ferreira, C.S. and Ferreira, A.J. (2001) Virulence properties of *Escherichia coli* isolated from ostriches with respiratory disease. Vet. Microbiol., 83: 71-80.

Kretzschmar, R., Barmettler, K., Grolimund, D., Yan, Y.-D, Borkovec, M., and Sticker, H. (1997) Experimental determination of colloid deposition rates and collision efficiencies in natural porous media. Water Res. Res., 33:1129-1137.

Kulabako N.R., Nalubega M., Thunvik R., (2008) Phosphorus transport in shallow ground-water in peri-urban Kampala, Uganda: Results from field and laboratory measure-ments. Environmental Geology, 53:1535-1551.

Kulabako, N.R., Nalubega, M., Thunvik, R. (2007) Study of the impact of land use and hydrogeological settings on the shallow groundwater quality in a peri-urban area of Kampala, Uganda .Science of the Total Environment, 381: 180–199.

Levy, J., Sun, K.,Findlay, R.H., Farruggia, F.T., Porter, J., Mumy, K.L., Tomaras, J.,and A. Tomaras, (2007) Transport of *Escherichia coli* bacteria through laboratory columns of glacial-outwash sediments: Estimating model parameter values based on sediment characteristics. Journal of Contaminant Hydrology, 89 : 71-106.

Li, X., Scheibe, T.D., and Johnson, W.P. (2004) Apparent decreases in colloid deposition rate coefficients with distance of transport under unfavorable deposition conditions: A general phenomenon. Environ. Sci. Technol., 38: 5616-5625.

Loveland, J.P., Bhattacharajeh, S., Ryan, J.N., and Elimelech, M. (2003) Colloid transport in geochemically heterogeneous medium: Aquifer tank experiment and modeling. Contaminant Transport, 65: 161-182.

Lutterodt, G, Basnet, M., Foppen, J.W.A. and Uhlenbrook, S. (2009a) The effect of surface characteristics on the transport of multiple *Escherichia coli* isolates in large scale columns of quartz sand. Water Research 43: 595-605

Lutterodt, G, Basnet, M., Foppen, J.W.A. and Uhlenbrook, S. (2009b) Determining minimum sticking efficiencies of six environmental *Escherichia coli* isolates. Journal of Contaminat Hydrology, 110: 110-117.

Lutterodt, G,., Foppen, J.W.A., Maksoud, A and Uhlenbrook, S. (2011) Transport of *Escherichia coli* in 25 m quartz sand columns. Journal of Contaminant Hydrology, 119: 80-88.

Macler, B.A., and Merkle, J.C. (2000) Current knowledge on groundwater microbial pathogens and their control. Hydrogeology Journal, 8: 29-40.

Madigan, M.T., Martinko, J.M., Dunlap, P.V. and D. P. Clark, (2009) Brock. Biology of Microorganisms. Twelfth Edition. Pearson Education Ltd. ISBN 0-321-53615-0.

Martin, M.J., Logan, B.E., Johnson, W.P., Jewett, D.G., and Arnold, R.G.(1996) Scaling bacteria filtration rates in different sized porous media. Journal of Env. Eng, 122: 407-415.

Matthess, G., Foster, S.S.D., and Skinner, A.C. (1985) Theoretical background, hydrogeology and practice of groundwater protection zones. IAH international contributions to Hydrogeology. Vol. 6.

Matthess, G., Pekdeger, A.,Schroeter,J. (1988) Persistence and transport of bacteria and viruses in groundwater- a conceptual evaluation. Journal of Contam. Hydrol., 171-188.

McCallou, D.R., Bales, R.C., and Arnold, R.G. (1995). Effect of temperature controlled motility on transport of bacteria and microspheres through saturated sediment. Water Resources Research, 31: 271-280.

Medema, G.J., Payment, P., Dufour, A., Robertson, W., Waite, M., Hunter, P., Kirby, R., Andersson, Y., (2003) Safe drinking water: an ongoing challenge. In: Dufour et al. (Eds.), Assessing Microbial Safety of Drinking Water: Improving Approaches and Methods, World Health Organization, ISBN 92 4154630.

Merkli, B (1975) Untersuchungen ueber Mechanismen und Kinetik der Elimination von Bakterien und Viren im Grundwasser. Diss. ETH Zurich 5420.

Morris, B. L., Lawrence, A. R. L., Chilton, P. J. C., Adams, B., Calow, R. C., and Klinck, B. A. (2003) Groundwater and its Susceptibility to Degradation. A Global assessment of the problem and options for management. Early Warning and Assessment Report Series, RS. 03-3.United Nations Environment Programme/DEWA, Nairobi, Kenya.

Morrow, J.B., Stratton, R., Yang, H.-H., Smets, B.F., and Grasso, D. (2005) Macro and nanoscale observations of adhesive behavior for several *E. coli* strains (0157:H7 and environmental isolates) on mineral surfaces. Environ. Sci. Technol., 39: 6395-6404.

Neihof, R., (1969) Microelectrophoresis apparatus employing palladium electrodes. J. Coll. Interface Sci., 30: 128–133.

Nield, D.A. and Kuznetsov, A.V. (2004) Forced convection in a helical pipe filled with a saturated porous medium. International Journal of Heat an Mass Transfer, 47: 5175-5180.

Nikaido H (2003). Molecular basis of bacterial outer membrane permeability revisited. Microbiol Mol Biol Rev, 67: 593–656.

Nyenje, P.M., Foppen, J.W., Uhlenbrook, S, Kulabako, R., and Muwanga, A (2010) Eutrophication and nutrient release in urban areas of sub-Saharan Africa — A review. Science of the Total Environment, 408, 447-455.

Ohshima, H (1994) Electrophoretic mobility of soft particles. Journal of Colloid and Interface Science, 163: 474-483.

Orskov, I, Orskov, F., Jann, B., and K. Jann, (1977) Serology, chemistry, and Genetics of O and K Antigens of *Escherichia coli*. Bacteriological Reviews,. 41 : 667-710.

Otto, K., Norbeck, J.., Larsson, T., Karlsson, K-A., and Hermansson, M. (2001) Adhesion of Type 1-Fimbriated *Escherichia coli* to Abiotic Surfaces Leads to Altered Composition of Outer Membrane Proteins. Journal of Bacteriology, 183: 2445–2453.

Owen, P., and Kaback, H. R. (1978) Molecular structure of membrane vesicles from *Escherichia coli*. Proc Natl Acad Sci USA 75: 3148–3152.

Ozeki, Y., Kurazono, T., Saito, A., Kishimoto, T., and Yamaguchi, M. (2003) A diffuse outbreak of enterohemorrhagic *Escherichia coli* O157:H7 related to the Japanese-style pickles in Saitama, Japan. Kansenshogaku Zasshi 77: 493–498.

Pang, L., Close, M., Goltz, M., Sinton, L., Davies, H., Hall, C., and Stanton, G. (2003) Estimation of septic tank setback distances based on transport of *E. coli* and F-RNA phages. Environ. Int., 29: 907-921.

Paramonova, E., Zerfoss, E.L., and Logan, B.E. (2006) Measurement of biocolloid collision efficiencies for granular activated carbon by use of a two-layer filtration model. Appl. Environ Microbiol., 72:5190-5196.

Pembrey, R.S., Marshall, K.C., and Schneider, R.P. (1999) Cell surface analysis techniques: What do cell preparation protocols do to cell surface properties? Appl Environ. Microbiol., 65: 2877-2894.

Powell, K.P., Taylor, R.G., Cronin, A.A., Barrett, M.H., Pedley, S., Sellwood, J., Trowsdale, S.A., and Lerner, D.N. (2003) Microbial contamination of two urban sandstone aquifers in the UK. Water Research, 37: 339-352.

Powelson, D.K. and Mills, A.L. (2001) Transport of *Escherichia coli* in sand columns with constant and changing water contents. Journal of Qual., 30: 238-245.

Pratt, L.A., and Kolter, R. (1998) Genetic analysis of *Escherichia coli* biofilm formation: Roles of flagella, motility, chemotaxis and type I pili. Mol. Microbiology, 30: 285-293.

Prigent-Combaret, C., O. Vidal, C. Dorel, and Lejeune, P. (1999) Abiotic Surface Sensing and Biofilm-Dependent Regulation of Gene Expression in *Escherichia coli*. Journal of Bacteriology, 181:5993–6002.

Redman, J., Grant, S.B., Olson, T.M., and Estes, M.K. (2001a). Pathogen filtration, heterogeneity, and potable reuse of wastewater. Environ. Sci. Technol. 35:1798-1805.

Redman, J.A, Estes, M.K. and Grant, S.B. (2001b). Resolving macroscale and microscale heterogeneity in virus filtration. Colloid and Surfaces A: Physicochemical and engineering Aspects, 191:57-70.

Ren, D., Bedzyk, L.A., Thomas, S.M. , Ye, R.W. ,Wood, T.K. (2004). Gene expression in *Escherichia coli* biofilms, Appl. Microbiol. Biotechnol., 64: 515–524.

Ren, Y, Liu, B., Cheng, J., Liu, F., Feng, L., and, Wang, L. (2008) Characterization of *Escherichia coli* O3 and O21 O antigen gene clusters and development of serogroup-specific PCR assays. Journal of Microbiological methods, 75: 329-334.

Rijnarts, H., Norde, W., Bouwer, E.J., Yklema, J.I., and Zehnder, A.J.B.(1993) Bacterial adhesion under static and dynamic conditions. Appl. Environ. Microbiol., 59: 3255-3265.

Salerno, M.B., Flamm, M., Logan, B.E., and Velegol, D. (2006) Transport of rodlike colloids through packed beds. Environ. Sci. Technol., 40: 6336-6340.

Sauer, K., (2003). The genomics and proteomics of biofilm formation. Genome Biology, 4:219.

Schembri, M.A., and. Klemm, P. (2001) Coordinate gene regulation by fimbriae-induced signal transduction. EMBO J, 20: 3074–3081.

Schembri, M.A., Kjærgaard, K., and Klemm, P. (2003) Global gene expression in *Escherichia coli* biofilms. Molecular Microbiology, 48 : 253 – 267.

Schijven, J (2001) Virus removal from groundwater by soil passage. Modeling field and laboratory experiments. Ph.D. Thesis. ISBN 90-646-4046-7. Posen and Looijen.Wageningen the Netherlands.

Schinner, T., Letzner, A., Liedtke, S., Castro, F.D., Eydelnant, I.A. and Tufenkji, N., (2010) Transport of selected bacteria pathogens in agricultural soil and quartz sand. Water Research, 44: 1182-1192.

Servin, A.L., (2005). Pathogenesis of Afa/Dr Diffusely Adhering *Escherichia coli*. Clinical Microbiology Reviews, 18: 264–292.

Shani, C., Weisbrod, N and Alexander, Y. (2008) Colloid transport through saturated sand columns: Influence of physical and chemical surface properties on deposition. Colloids and Surfaces. A: Physicochem Eng. Aspects, 316: 142-150.

Sharma, M.M., Chang, Y.I., and Yen, T.F. (1985). Reversible and irreversible surface charge modification of bacteria for facilitating transport through porous media. Colloid Surf., 16:193-206.

Shellenberger, K., and Logan, B.E. (2002) Effect of molecular scale roughness of glass beads on bacterial and colloidal deposition. Environ. Sci. Technol., 36: 184-189.

Shemarova, I.V., and Nesterov, V.P. (2005) Evolution of mechanisms of Ca^2+-signalling: Role of Calcium ions in signal transduction in prokaryotes. J. Evol. Biochem and Physiol., 41:12-19.

Simoni, S.F., Harms, H., Bosma, T.N.P. and Zehnder, A.J.B. (1998) Population heterogeneity affects transport of bacteria through sand columns at low flow rates. Environ Sci and Technol., 32: 2100-2105.

Šimůnek, J., Šejna, M., Saito, H., Sakai, M and van Genuchten. M, T (2008) The HYDRUS-1D Software Package for Simulating the One-Dimensional Movement of Water, Heat, and Multiple Solutes in Variably-Saturated Media Version 4 Department of Environmental Sciences University of California Riverside, Riverside California.

Šimůnek, J., Šejna, M., Saito, H., Sakai, M., and Th. van Genuchten M. (2009) The HYDRUS-1D Software Package for Simulating the One-Dimensional Movement of Water, Heat, and Multiple Solutes in Variably-Saturated Media Version 4.08. Department of Environmental Sciences University of California Riverside, Riverside California.

Sinton, L.W., Finaly, R.K., Pang L and Close, M. (2000) Transport and attenuation of bacteria and bacteriophages in alluvial gravel aquifer. New Zealand Journal of Marine and Freshwater Research, 34:175-186

Sinton, L.W., Noonman, M.J., Finaly, R.K., Pang L and Scott, D.M. (1997) Transport of bacteria and bacteriophages in irrigated effluent into and through an alluvial aquifer. Water, Air and Soil Pollution, 98: 17-42.

SPSS (2005). Software Package for Social Sciences, Release 14.0.0. 2005. Chicago: SPSS Inc.

Stenutz, R., A. Weintraub, and Widmalm, G. (2006) The structures of *Escherichia coli* O-polysaccharide antigens. FEMS Microbiol Rev, 30 : 382–403.

Tabe Eko Niba, E., Naka, Y., Nagase,M.,. Mori, H and Kitakawa, M. (2007) A Genome-wide Approach to Identify the Genes Involved in Biofilm Formation in *E. coli*. DNA Research pp. 1–10.

Tarr, P.I., Bilge, S.S. , Vary, jr., J.C. , Jelacic, S., Habeeb, R.L., Ward, T.R. , Baylor, M.R. , and Besser, T. E (2000) Iha: a Novel *Escherichia coli* O157:H7 Adherence-Conferring Molecule Encoded on a Recently Acquired Chromosomal Island of Conserved Structure. Infection and Immunity 68: 1400–1407.

Taylor, R., Cronin, A., Pedley, S., Barker, J., and Atkinson, T. (2004) The implication of groundwater velocity variations on microbial transport and well head protection-review of field evidence. FEMS Microbiol. Ecology, 49: 17-26.

Tong, M., and Johnson, W. (2007) Colloid population heterogeneity drives hyper-exponential deviation from classic filtration theory. Environ. Sci. Technol., 41: 493-499.

Tufenkji, N., (2006) Application of a dual deposition mode model to evaluate transport of *Escherichia coli* D21 in porous media. Water Resour. Res., 42, W12S11, doi: 10.1029/2005WR004851.

Tufenkji, N., and Elimelech, M (2005a) Breakdown of colloid filtration theory: Role of the secondary energy minimum and surface charge heterogeneities. Langmuir, 21: 841-852.

Tufenkji, N., and Elimelech, M. (2004a) Correlation equation for predicting Single-collector efficiency in physicochemical filtration in saturated porous media. Environ. Sci. Technol., 38: 529-536.

Tufenkji, N., and Elimelech, M. (2004b) Deviation from the classical colloid filtration theory in the presence of repulsive DLVO interactions. Langmuir, 20:10818-10828

Tufenkji, N., and Elimelech, M. (2005b) Spatial distribution of *Cryptosporidium oocysts* in porous media: Evidence of dual mode deposition. Environ. Sci. Technol., 39:3620-3629.

Tufenkji, N., Redman, J.A. and Elimelech, M. (2003) Interpreting deposition patterns of microbial particles in laboratory-scale column experiments Environ. Sci. Technol. 37:616-623.

Ulett, G.C., Webb, R.I. and Schembri, M.A. (2006) Antigen-43 –mediated autoaggregation impairs motility in *Escherichia coli*. Microbiology, 152: 2101-2110.

van der Mei, H.C., and Busscher, H.J.(2001) Minireview: Electrophoretic mobility distributions of single-strain microbial populations. Appl. Env. Microbiol., 67: 491-494.

van der Wal, A., Minor, M., Norde,W., Zehnder, A. J. B and Lyklema, J. (1997) Electrokinetic Potential of Bacterial Cells. Langmuir, 13:165-171.

van Houdt, R., and Michiels, C.W. (2005) Role of bacterial cell surface structures in *Escherichia coli* biofilm formation. Research in Microbiology 156 : 626–633.

van Loosdrecht, M., Lyklema, J., Norde, W., Schraa,G., and Zehnder, A. (1989) Bacteria adhesion: a physicochemical approach. Microbiol Ecology, 15: 1-15.

van Loosdrecht, M.C.M., Lyklema, J., Norde, W., Schraa, G., and Zehnder,A.J.B. (1987a) The role of cell wall hydrophobicity in adhesion. Appl. Environ. Microbiol., 53: 1893-1897.

van Loosdrecht, M.C.M., Lyklema.J., Norde, W., Schraa, G., and Zehnder, A.J.B. (1987b) Electrophoretic mobility and hydrophobicity as a measure to predict the initial steps of bacterial adhesion. Appl. Environ. Microbiol., 53: 1898-1901.

Velasco-Casal, P., Wick, L.Y., Ortega-Calvo, J.J., (2008) Chemoeffectors decrease the deposition of chemotactic bacteria during transport in porous media. Environ. Sci. Technol., 42: 1131-1137.

Walker, S.L,. Redman, J.A, and Elimelech, M (2004) Role of Cell Surface Lipopolysaccharides in *Escherichia coli* K12 Adhesion and Transport Langmuir, 20: 7736-7746.

Walker, S.L., Hill, J.E., Redman, J.A. and Elimelech, M. (2005) Influence of growth phase on adhesion kinetics of *Escherichia coli* D21g. App Env Microbiol., 72:3093-3099.

Walker, S.L., Redman, J.A., and Elimelech, M. (2005) Influence of growth phase on bacterial deposition: Interaction mechanisms in packed-bed column and radial stagnation point flow systems. Environ. Sci. Technol., 39: 6405-6411.

Weiss, T.H., Mills, A.L., Hornberger, G.M., and Herman, J.S. (1995) The effect of bacterial cell shape on transport of bacteria in porous media. Environ. Sci. Technol., 29: 1737-1740.

World Health Organization, (2004) World Health Report. http://www.who.int/whr/2004/en/report04 en.pdf

Yang, H.-H. (2005) The effect of environmental stress on cell surface properties and their relation to microbial adhesion in feedlot *Escherichia coli* isolates Ph.D. dissertation University of Connecticut 2005.

Yang, H-H, Morrow, J.B, Grasso, D, Vinopal, R.T, Dechesne A and Smets B (2008) Antecedent Growth Conditions Alter Retention of Environmental *Escherichia coli* Isolates in Transiently Wetted Porous Media Environmenttal Science and. Technology, 42: 9310–9316.

Yang, H-H., Morrow, J.B., Grasso, D., Vinopal, R.T., and Smets, B.F. (2006) Intestinal versus external growth conditions change the surfacial properties in a collection of environmental *Escherichia coli* isolates. Environ. Sci. Technol., 40: 6976-6982.

Yang, H-H., Vinopal, R.T., Grasso, D., and Smets, B.F. (2004) High diversity among environmental *Escherichia coli* isolates from a bovine feedlot. Appl. Environ. Microbiol., 73: 1528-1536.

Yao, K., Habibian, M.T. and O'Melia, C.R. (1971) Water and Waste Filtration: Concepts and Applications. Environ Sci Technol., 5.1105-1112.

Samenvatting

Doordat depositie van bacteriën op korrels in poreuze media afwijkt van wat theoretisch voorspeld wordt, is het tot nu toe onmogelijk gebleken om het transport van bacteriën in aquifers nauwkeurig te voorspellen. In het geval van pathogene bacteriën kan dit gebrek aan kennis uiteindelijk leiden tot het onbedoeld vervuilen van drinkwatervoorzieningen (bronnen, geboorde of gegraven putten), als bijvoorbeeld sprake is van faecale verontreinigingen. Vanwege het belang van de *Escherichia coli* (*E. coli*) als een indicator voor fecale verontreiniging van het drinkwater, richt dit proefschrift zich op het transport van *E. coli* in verzadigde poreuze media. De doelstellingen waren om (i) hechtingsverschillen te bestuderen van een enkele *E. coli* stam, maar ook hechtingsverschillen tussen meerdere E. coli stammen, (ii) variatie in hechtingsverschillen tussen cel populaties te karakteriseren, (iii) een methode te ontwikkelen om de minimale hechtingsefficiency te bepalen en om (iv) de bijdrage van verschillende cel-eigenschappen op bacteriële hechting aan sedimentaire kwarts korrels te bepalen.
Het grootste deel van dit onderzoek is uitgevoerd onder laboratorium omstandigheden (bijvoorbeeld de kolom experimenten en zgn. batch experimenten). Een ander deel van dit onderzoek richtte zich op de transport eigenschappen van *E. coli* stammen geïsoleerd uit eindpunten van de grondwaterstroomlijnen (bronnen) in Kampala, Uganda. De onderliggende hypothese was dat het transport van een dergelijke groep *E. coli* stammen mogelijk kan worden gekenmerkt door een zelfde set van transport parameters.

Het transport van *E. coli* stammen geïsoleerd uit verschillende milieus is bestudeerd met behulp van verzadigde kwarts zand kolommen. Korte (7 cm) en lange (1.5 - 25 m) kolommen met multiple sampling ports zijn gebruikt om de hechtingsvariatie tussen verschillende *E. coli* stammen te bestuderen, om de variatie binnen een stam beter te karakteriseren en om een methode te ontwikkelen om de minimale hechtingsefficiency van E. coli te bepalen. De zeer lange kolommen (25 m) zijn gebruikt om de minimale hechtingsefficiency ook daadwerkelijk te meten. Ook werden de fenotypische kenmerken en genen, die coderen voor structuren aan de buitenste *E. coli* cel membraan, bepaald om een eventueel verband te leggen met het transport deze *E. coli* cellen.

Geen van de bestudeerde *E. coli* cel karakteristieken had een significante invloed op de hechting van E. coli aan kwartskorrels. Echter, cel motiliteit en antigen-43 expressie bevorderden hechting over relatief korte transport afstanden. Een substantieel deel van de *E. coli* isolaten uit de grondwater bronnen van Kampala behoorde tot hetzelfde serotype (*E. coli* O21: H7). Op basis hiervan concluderen wij dat stammen van dit serotype kennelijk bepaalde kenmerken hebben, die hun selectieve transport door de aquifers van Kampala bevorderen en die wellicht de geobserveerde uniformiteit in transport parameters kunnen verklaren. Met uitzondering van de Kampala *E. coli* isolaten, laten de resultaten vooral zien, dat er sprake is van grote hechtings heterogeniteiten, niet alleen van verschillende *E. coli* stammen, maar zelfs ook binnen een en dezelfde stam.

We hebben praktisch relevante lage waarden van hechtingsefficiencies gemeten bij zeer lange transportafstanden en daarmee het belang aangetoond van het gebruik van lange kolommen. De gemeten lage waarden van de hechtingsefficiency geven aan, dat voor bacteriële populaties gelekt naar grondwater, bepaalde sub-populaties zgn. non-attaching kenmerken kunnen bezitten, waardoor de kans groot is, dat deze sub-populaties over grote afstanden getransporteerd kunnen worden. Bestaande drinkwater beschermings regulering onderschat dergelijke grote transport afstanden.

Variaties in hechtingsefficiency binnen een en dezelfde *E. coli* stam kan worden uitgedrukt middels een zgn. power-law verdeling tussen de fractie van cellen als funcite van hun hechtingsefficiency. De minimale hechtingsefficiency kan worden afgeleid van deze power-law verdeling. De minimale hechtingsefficiëntie is door ons gedefinieerd als de hechtingsefficiency behorend bij een bacterie fractie van 0,001% van de initieel geinjecteerde hoeveelheid bacteriën, na verwijdering van 99,999% (5 log reductie) van de oorspronkelijke hoeveelheid bacteriën. De op deze wijze verkregen waarden waren lager dan die Geëxtrapoleerde waarden waren lager dan gemeten met experimenten en zijn een waardevol hulpmiddel bij het bepalen van een goed grondwater beschermingsgebied in real-world scenario's.

Toekomstig onderzoek zou zich moeten richten op structuren aan het oppervlak van *E. coli* cellen, die betrokken zijn bij de eerste hechting van *E. coli* cellen op weefsels en / of abiotische oppervlakken. Verder denken we dat het belangrijk is om meer experimenteel werk met bacterie transport in het veld uit te voeren in plaats van in het laboratorium, omdat het interpreteren van laboratorium onderzoek ernstig beperkt wordt door transport afhankelijke schaal problematiek

Curriculum Vitae

George Lutterodt was born in the early 1970s at Osu in Accra, Ghana. He graduated from the University of Ghana and the Royal Institute of Technology (KTH) in Stockholm, Sweden with a BSc in Geology and an MSc in Environmental Engineering and Sustainable Infrastructure, respectively, in 1998 and 2004. Between the period 2005 and 2006, he worked with Nii Consult, a Water Resources and Environmental Management Consultancy, and also with Geogroup Limited as a Field Hydrogeologist. Prior to his post-graduate studies and during the period 1998-1999 he worked in the Groundwater Division of the Water Research Institute of the Council for Scienific and Industrial Research in Ghana as a National Service Person. Between the years 2000 to 2005 he was employed as an Assistant Geologist at the Ghana Geological Survey Department, he also took a teaching job at the Ghanata Secondary School in Dodowa, Accra and taught High School Physics and mathematics from 2000-2002. From January 2007 to May 2011, he studied in Delft for his PhD research.

T - #0155 - 160425 - C6 - 240/166/8 - PB - 9780415621045 - Gloss Lamination